PRACTICAL PROCESS CONTROL

PRACTICAL PROCESS CONTROL

A M Seal
Consultant Process Control Engineer, UK

A member of the Hodder Headline Group
LONDON • SYDNEY • AUCKLAND

Copublished in North, South and Central America by
John Wiley & Sons Inc., New York • Toronto

First published in Great Britain 1998 by Arnold,
a member of the Hodder Headline Group,
338 Euston Road, London NW1 3BH
http://www.arnoldpublishers.com
Copublished in North, Central and South America by
John Wiley & Sons, Inc., 605 Third Avenue, New York, NY 10158 0012

Whilst the advice and information in this book is believed to be true and
accurate at the date of going to press, neither the author nor the publisher
can accept any legal responsibility or liability for any errors or omissions
that may be made.

British Library Cataloguing in Publication Data
A catalogue record for this book is available from the British Library

Library of Congress Cataloging-in-Publication Data
A catalog record for this book is available from the Library of Congress

ISBN 0 340 70590 6
ISBN 0 470 28321 1 (Wiley)

Publisher: Matthew Flynn
Production Editor: James Rabson
Production Controller: Priya Gohil
Cover designer: Terry Griffiths

Typeset in 10/12pt Times by Anneset
Printed and bound in Great Britain by J. W. Arrowsmith Ltd, Bristol

The Author

The author served an Electrical and Mechanical Student Apprenticeship during the early 1960s. Since completion of his apprenticeship he first worked in the telecommunications industry and subsequently in instrumentation and process control for over 26 years covering many aspects of the application and maintenance of control equipment. In the early years of working in instrumentation and control most of the equipment was pneumatic, and with this, and early electronic instrumentation, a great amount of experience was gained in maintenance, installation and commissioning control equipment applied to many different process industries. Since those early beginnings he was fortunate to have been at the forefront of the application of new technology as it became available. Initially starting as an Instrument Technician his working roles have included process plant shift maintenance, oil refinery shut-down maintenance, controls for boilers and turbines in the marine industry, chemical process controls and all aspects of paper production controls.

For the last 20 years he has worked in the paper industry and has been responsible for the introduction and application of modern control techniques, including Distributed Control Systems, into many plants. His role as Group Process Control Engineer with a leading UK paper company has involved him in the automation of a large amount of process plant, involving specification, selection, installation and application of the latest technology control equipment as it has become available. Much of the work has involved interfacing various types of control systems equipment and overcoming the resulting problems. Working as a member of project teams has generated an awareness of a shortfall of understanding by many engineers not familiar with process control and not sure of 'how and what' to select for control purposes. The result is this book which is intended to give a guide to such engineers faced with selection of control equipment, to enable a much better identification of the correct equipment needed for projects involving process control.

In his current post his work covers many of the UK pulp, paper and board manufacturers, plus other varied industries, covering all aspects of associated process control and instrument application. Repeatedly he finds situations where a basic understanding of what is required for good process control is often missing, primarily due to the severe shortage of available, suitable, experienced and qualified personnel. Having identified the situation, this book uses the author's experiences to attempt to bridge the gap, thus enabling those not fortunate enough to have all the necessary skills to choose suitable process control equipment.

Contents

Acknowledgements

Thanks must go to the many companies listed alphabetically below who have willingly supplied much appreciated information, photographs, drawings, etc. Details of their major offices are listed, should any reader wish to contact them for more information. It has been a difficult but enjoyable job to select the data from the overwhelming amount provided. Most thanks must, however go to my wife, Mary, for her encouragement, patience and support for the project.

Camille Bauer Controls Ltd
Priest House
Cradley Heath
Warley
West Midlands B64 6JN
UK
☎ 01384 638822

Camille Bauer-Metrawatt AG
Aagauerstrasse 7
CH-5610 Wohlen
Switzerland
☎ +41 56 618 21 11

Croft Instruments Ltd
The Croft
Park Lane
Carhampton
Minehead
Somerset TN24 6NN
UK
☎ 01643 821530

Coulton Instrumentation Ltd
17 Somerfield Business Park
Christchurch
Dorset BH23 3RU
UK
☎ 01202 480303

Danfoss Instruments Ltd
Perivale Ind. Park
Horsenden Lane South
Greenford
Middlesex UB6 7QE
UK
☎ 0181 991 7000

DeZurik International Ltd
Nelson Way
Cramlington
Northumberland NE23 9BJ
UK
☎ 01670 714111

Endress+Hauser Ltd
Floats Road
Manchester M23 9NF
UK
☎ 0161 286 5000

Fisher-Rosemount Control Systems
Meridian East
Leicester LE3 2WU
UK
☎ 0116 2822822

Fisher-Rosemount Control Valves
Horsfield Way
Bredbury Industrial Estate
Stockport
Cheshire SK6 2SU
UK
☎ 0161 406 8826

Fisher Rosemount Measurements
Heath Place
Bognor Regis
West Sussex PO22 9SH
UK
☎ 01243 863121

Fisher Controls International Inc
P.O. Box 190
Marshalltown
Iowa 50158
USA
☎ (515) 754 3000

Honeywell Control Systems Ltd
Block 1A
Newhouse Industrial Estate
Mortherwell
Lanarkshire ML1 5SB
UK
☎ 01698 481000

Honeywell Controls Ltd
Honeywell House
Arlington Business Park
Bracknell
Berkshire
UK

Honeywell Industrial Instrumentation & Control
16404 North Black Canyon Hwy
Phoenix AZ 85023
USA

Kilkie Paper Mill Services
Melville Square
Comrie
Perthshire PH6 2DL
UK
☎ 01764 670141

Krohne Ltd
Rutherford Drive
Park Farm Industrial Estate
Wellingborough
Northamptonshire NN8 6AE
UK
☎ 01933 408500

Krohne Messtechnik GmbH & Co. KG
Pastfach 100862
D-47008 Duisburg
Germany
☎ (0203) 301-0

Krohne America Inc
7 Dearborn Road
Peabody MA 01960
USA
☎ (508) 535-6060

Millwide Engineering Services Ltd
Ewood Bridge Mills
Manchester Road
Ewood Bridge
Haslingden
Rossendale
Lancashire BB4 6LD
UK
☎ 01706 222203

Moore Products Co (UK) Ltd
Copse Road
Lufton
Yeovil
Somerset BA22 8RN
UK
☎ 01935 706262

Moore Products Co
Sumneytown Pike
Spring House PA 19477
USA
☎ (215) 646-7400

Siemens plc
Sir William Siemens House
Princess Road
Manchester M20 2UR
UK
☎ 0161 446 5000

Spirax-Sarco Ltd
Charlton House
Cheltenham
Glocestershire GL53 8ER
UK
☎ 01242 521361

VEGA Controls Ltd
Kendal House
Victoria Way
Burgess Hill
West Sussex RH15 9NF
UK
☎ 01444 870055

VEGA Controls
Ohmart Corp
4141 Allendorf Drive
Cincinnati
Ohio 45209
USA
☎ 05 13 272 0131

Introduction

The purpose of this book is to introduce process control to those engineers and technicians who are unfamiliar with process control techniques, and who have varying needs to understand how to apply controls to process. It is also intended to assist those engineers who have been involved with process control in the past, and are not so familiar with the type and range of control equipment available today. The aim is to keep the information general, applicable to most processes requiring varying degrees of control, to identify what specific information is required by suppliers, and to identify the most suitable equipment for the application in hand. Additionally a degree of troubleshooting will be covered, as well as system design, installation, commissioning and running plant tuning. In many sections there are guidelines on how to install process control equipment, especially measurement devices. To help overcome many of the problems generated by poor installations a few examples of how not to install are included.

The reasons for applying process control to any plant fall into two basic categories: first to improve, or establish, management information relating to the process, and second to apply, or improve, control in an effort to improve plant efficiency, or reduce manning levels. The plant to which process control is applied can also be split into two general types, the first being a totally new plant and the other an existing plant. The new plant offers the opportunity to apply the latest technology, carefully selected, to suit the application. With current day requirements for good control, good operator interfacing and management information there is a large selection of equipment to do the job. To ensure that the most suitable equipment is chosen the system specifications detailed in this book apply equally to new plant, as well as to up-grades of existing plant. The application to existing plant may be anything from a small amount, say one control loop, up to a major plant up-grade. Where existing plant is to be up-graded there is a high probability that some instrumentation and control has already been applied, and it should be taken into consideration when evaluating the project needs. If application of control has been on an *ad-hoc* basis there is every chance that a careful survey will identify the equipment which has given reliable service and that which has been troublesome. Continuing application of known unreliable equipment will not be appreciated by those who will have to maintain it, so a little research of equipment performance will pay dividends.

Today there are numerous manufacturers making many different types of control equipment. This can present a daunting problem of choice to those with little experience of how to, and what is needed to, control the process. Many manufacturers align their equipment to specific industries, for reasons of design, success record and many other factors. Hopefully this book will enable the reader to be better able to evaluate the suitability of equipment for the job in hand, and to generate a specification for the items required.

Several of the major control equipment manufacturers, and suppliers, have published handbooks and guide books introducing 'process control'. These books are very informative, but we must remember that they are provided primarily to promote sales of their equipment. These publications are nevertheless well recommended and, if acquired, should be made part of your reference library. However, this book is aimed at giving you the ability to make the choice of the best equipment, of whatever manufacture, and the best application of such equipment for the job, wherever possible, hopefully without bias to any particular manufacturer. Many suppliers have been extremely helpful, and have contributed information, pictures, etc., which has been greatly appreciated, but I have tried to be unbiased and have selected data which I felt would be of assistance in identifying what equipment looks like, and showing how it fits together. The information included is a small selection of what is available on the market today, but hopefully will give you some idea of where to start.

All the information in this book is based on over 25 years of working in various industries using instrumentation and process control equipment applied to many different processes. It covers equipment specification, system design, equipment selection, installation, commissioning and a brief outline of control loop tuning.

Some of the information may seem very basic, just common sense, but so often the basics are not transmitted to the potential supplier, or installer, with the result that the wrong equipment is supplied, or worse! Ensuring that even the basics are covered, as well as the important detail, will ensure a good and safe project, along with a good customer–supplier relationship.

The golden rule we should always remember when specifying any equipment is that '*you get what you pay for, and if it is not specified, it will not be included*'.

It is very important that in the application of any equipment to plant or process where any special conditions are applicable, e.g. **hazardous area classification**, the applicable *regulations and guidelines must be observed* and fully detailed to all potential equipment suppliers.

To get a better understanding of how to specify and select instrumentation and process control equipment, and appreciate what potential suppliers are proposing, it is worth understanding some of the terminology associated with instrumentation and process control. The terminology covered here is by no means definitive, but should give an insight to the majority of terms you are likely to encounter. With all electronic equipment where microprocessors are used there is an ever growing amount of jargon which also impacts on instrumentation and process control equipment. Some has been included, but it is impossible to keep up with this growth, so please forgive any shortfall, especially in this area.

Process control terminology

Control device: also called a **final control element**, can take several forms and is the device to act on the process parameter to achieve control of that parameter.

Controller: the unit which manipulates a *control device* to maintain the process parameter under control, at *set point*, as a result of variations presented to the unit by a *measuring device*.

Demand: similar to *load change* and is the requirement of the controller on the process parameter to maintain, or regain, control at *set point*, after a change made by operator or system.

Deviation or error: the offset of the process parameter from *set point*. This should not be confused with *offset*.

Flow switches: similar action to *pressure switches* below, but generally applicable to fluid flows, and use several different types of operating principles according to the type of application.

Level switches: very similar to *pressure switches* below, but switching on liquid level, etc., e.g. float switch.

Live zero: the elevated signal output representing the calibrated zero of the transmitter. The live zero enables an assessment of any process value which is below the calibrated zero, and hence an exact zero calibration point of the measured variable.

Load change: the change effected by the controller to maintain the process parameter at *set point* as a result of a *process upset* or disturbance.

Load response: the way a process reacts after a deviation in the *process load*.

Measuring device: generally called a **transmitter**, also known as a sensor and sometimes inaccurately termed a meter, measures a process parameter and generates a signal suitable for use by control equipment from this measurement. A **transmitter** is an analogue device which converts a measured variable into a signal of standard format, e.g. 4 to 20 mA, 3 to 15 psig, 1 to 5 V, etc. There are some digital output transmitters available now, although these generally employ HART™ communication protocol, superimposed on a standard 4 to 20 mA analogue signal. Various types of transmitters are described in Chapter 2 on measurement devices.

Offset: the constant deviation from *set point* when the process parameter is under steady state control.

Pressure switches: unlike transmitters these devices will only give a change of switch state, at a preset point, relative to a measured variable. Due to their design there will always be a slight 'dead band' between the rising and falling switching point. This must be taken into account when specifying the point at which the switch is to operate, i.e. rising or falling.

Process: this refers to the physical process and action performed, chemical reaction, or the collection of equipment making up the plant to carry out a process.

Process lag: the real time taken for the transmitter to identify a change made by the *control device*. If the change is seen very quickly the process is termed 'fast', and if the time taken is many seconds or minutes the process is termed 'slow'. This is also known as **dead time** and should not be confused with *load response*.

Process load: the level of force, material, power, etc. applied or removed from the process under control.

Process upset: the disturbance of the process parameter away from the *set point*, detected by the *measuring device*/transmitter.

Set point: this is the desired running/operating value of the process parameter under control.

Temperature switches: very similar to *pressure switches* above, usually referred to as **thermostats** when applicable to domestic heating systems etc.

Transducer: similar to a transmitter but generally generates a signal which is not one of the recognised standards, e.g. 4 to 20 mA and requires additional signal conversion to generate a standard signal.

Zero suppression/elevation: on many applications the working zero point may not be the physical zero encountered, e.g. pressure transmitter connected to a steam pipe which is 3 m above the transmitter will require a zero suppression of 3 m WG to read 0.0 bar when the pressure is zero.

There are many other terms which you will encounter, but these are not so common as the above and are generally self-explanatory.

We now need to define some of the control actions in order to understand what 'control' facilities are required:

Derivative (D) or Rate control: is a control action which responds to the rate of change of the controlled process parameter. It must not be used on its own as it cannot re-establish control at *set point* on its own. Its prime function is to offset the instability which can be introduced by the other control actions, but it is quite often applied badly giving rise to its not being used for fear of causing problems! In many general applications, however, it is not necessary, and hence its value is not exploited where it can be very useful.

Integral (I) or Reset control: a control action on the error between *set point* and process parameter to reduce this error to zero. Like *Proportional control* it responds to magnitude and direction. It is rarely used on its own. Integral is generally expressed as seconds, or minutes, per repeat and reset as repeats per minute (reciprocal of integral).

Proportional (P) or Gain control: also known as **Throttling control**, this is the controller response to the magnitude and direction of a change in the process parameter under control. The proportion setting is expressed as % and the Gain is the reciprocal of the proportion % as a finite figure (i.e. gain = 1 = Proportion setting = 100%, Gain = 2 = Proportion setting = 50%).

From the above we can see where the term **PID control** comes from: it is the combination of all the above actions in one *controller*. The actions can be turned on, or off, as required. You may also find the term PI used for controllers with Proportional and Integral

Control action: the relationship of the controller output and the measured variable input and will be either **direct action** (rise in process parameter above *set point* resulting in rising controller output signal), or **reverse action** (rise in process parameter above *set point* resulting in falling controller output signal).

We also have some other control actions, strategy or functions which it is worth defining.

Batch control: a control system to suit a 'batch' type process where there are distinct intervals between completion of each process batch and commencement of the next. The time interval between batches can be used for recipe change, etc.

Cascade control: a control system where a master controller, in control of a process parameter, generates the *set point* of a controller of another 'faster' parameter which affects the master parameter. In this system the master must always be the 'slower' process parameter, otherwise uncontrollable instability will result. A good example of cascade control is the control of boiler water level, the level control being the master, generating the *set point* for the cascaded feedwater flow control.

Dead band: a period of 'freeze' of *controller* action, or a small band between turning 'on' and turning 'off' a switch, such as a *level switch*.

Lead/lag: a form of control normally associated with *ratio* or *cascade control* where one process parameter is required to 'lead' the other in one direction, but 'lag' in the other. This type of control is to be found in burner fuel/air control, e.g. on oil fired burners the air flow must lead the oil flow on rising firing rate, and lag the oil flow on falling firing rate, to avoid the generation of smoke.

Ratio control: a control combination which gives the facility to 'blend' two or more flows, or other parameters, together in a process where a master parameter will generate the *set point* for a slave process parameter controller, or a master controller will generate the *set points* for the blended parameter controllers.

Selective control or auto-selector control: a control system where several controllers manipulate one *control device* by means of a *signal selector* (high, low or median) in order to maintain control of the process, or keep a process parameter within specific limits.

Signal selector: a device which selects one control signal to pass to the selector's output according to its state relative to the selector function, from two or more input signals, e.g. a 'low signal selector' will pass the lowest of the input signals to the output.

Now we cover some of the other terminology which will be encountered in this book or in process control application.

CAD: Computer Aided Design. To the layman this is a computerised design drawing system.

CIM: Computer Integrated Manufacturing. This will normally be associated with other computer functions such as CAD, or a master computer system generating production recipes, e.g. *batch control*.

DCS: Distributed Control System. See Chapter 7 for more detail.

HART^{TM}: Highway Addressable Remote Transducer. This is a specialised electronic communications protocol for transmitters, enabling information and calibration to be transmitted over the signal cables.

Loop diagram: a line diagram showing the linking of all the components included in a control loop.

MCC: is a Motor Control Centre and is similar to an MMP which is a Multi-Motor (control) Panel.

MES: Manufacturing Execution System. Not covered in any detail in this book. This is a high level computer system, interfaced to factory floor control systems for download of recipes, production requirements, etc.

MIS: Management Information System. Not covered in any detail in this book. It is a high level integration of all workplace computer systems giving information to the high level computer of production, raw material status, finished goods stock, etc.

P & I Diagram: a Process and Instrument Diagram, this is a *process flow sheet* with most, or all, of the instrumentation and control equipment shown on the drawing using standard symbols.

PC: Personal Computer. Industrial PCs are usually 'hardened' to make them more suitable for the industrial environment. This usually entails fitting the electronics into dust proof enclosures, and installing the disk drives inside the cabinet.

PLC: Programmable Logic Controller. See Chapter 6 for more detail.

Process flow sheet: a line drawing giving a diagrammatic representation of the process using standard symbols, etc.

PSU: Power Supply Unit.

PV: Process Variable.

SCADA: Supervisory Control And Data Acquisition. The term given to the operator interface to a *PLC* network. Most SCADA systems employ *PCs* loaded with special software which is programmed to customise it to the process under control.

SI: Systems Integrator. A rare, costly person, highly skilled and experienced, who is capable of ensuring the compatibility of equipment and can ensure that all items will work together or sort out the problems encountered when systems do not perform correctly when interfacing on the same communications network.

Windows NT: trademark for Microsoft Windows New Technology, where Windows software is applied to industrial applications.

Other terminology and abbreviations used in this book will be defined wherever they first appear in the text.

1 What type of process — what type of control?

The first stage to be undertaken, before specifying any equipment, is to identify *what type of process* we are trying to control. Establishing which type of process is to be controlled will reduce the amount of equipment we need to consider, and give a better understanding of what needs to be considered when selecting equipment, as well as a better appreciation of the process itself. First we should ask ourselves a basic question: Is the process **continuous** or **cyclic**? The answer will generally determine what type of control is required. Cyclic control is also known as **on/off** control, which possibly best describes the control type. Continuous control, sometimes known as **analogue control**, is, as the title describes, a control which continually modulates something to keep a process under control. Continuous control can be sub-divided into several control formats, which we will cover later in this book. With any control application we need to identify the major control requirements of the process, to make the selection of control equipment which best suits the need. Identifying the type of control can drastically reduce the equipment you need to consider, and consequently will save time and effort and hence cost. With many selection processes it is helpful to attempt to categorise requirements in order to make life simple. In the case of different control applications, however, it may not be very easy to categorise requirements, and in all cases categorisation should be carefully undertaken, to avoid jumping to wrong conclusions, and hence any potential error in identifying requirements. Some control applications can be very grey as to recognition of their control type so over the following pages hopefully we will clarify the information needed to make an accurate assessment of requirements of all control equipment, in all areas of process control application, and hence categorise the control aspects correctly.

1.1   Process flow sheet

The starting point for application of any control equipment is the **process flow sheet**. The flow sheet should show all items of process plant, e.g. pumps, tanks, valves, etc., and is a line diagram of the process itself. The drawing of the flow sheet, or **flow diagram**, is normally the responsibility of the process designer, or design team. If it is available it will give much of the information needed to identify the type of process, and can be used as the skeleton for the **P & I (Process & Instrumentation) diagram**. If it is not available then drawing this should be the first task undertaken, since it will form the basic communication and reference document for the project. Request the designer to provide the flow sheet, along with a written description if possible. The designer will know how the process is expected to function, and the controls can be added to ensure that functionality is obtained. Although some control requirements may be obvious, there may be interactions or requirements which are not clear. The addition of instrumentation installed in the field (on plant), including **transmitters**, **control valves** and other items used for control, to the process flow sheet to form the P & I diagram will give a picture of the process, and what information is available, measured and controlled on the process. The generation of a process operation description to be read in conjunction with the process flow sheet will be an added bonus, and should start to give a clear identification of whether the process is mainly cyclic or continuous. The P & I diagram should show which **transmitter** links to which controller, and then to which **final control device**. This will be defined by the process operation description, and in essence should be a pictorial representation of that description. The following drawings of flow diagrams for a few control loops will give a guide to which type of control is most suited to the application.

1.2   Loop diagram

For each parameter under control there will be a number of items linked together to achieve the control necessary for that parameter. In some cases one piece of control equipment may comprise more than one control item, e.g. a ball-float valve (ball-cock) which is a complete level control loop (level controller) in one package. If you are not familiar with the make-up of the items required to satisfy the job in hand, it is always worthwhile drawing a simple loop diagram for each control function. This will help to identify the type of control items/elements required. As you consider the elements that make up the loop, the diagram can be developed. This diagram can then form part of the project documentation for installation, and on-going maintenance. It will also identify any additions necessary to the P & I diagram which is effectively the process flow sheet with the control loops superimposed on it. It is essential that the **measurement**, and **con-**

trol devices, identified for all control loops, are also identified on the P & I diagram. One should be conformation, and back check, of the other.

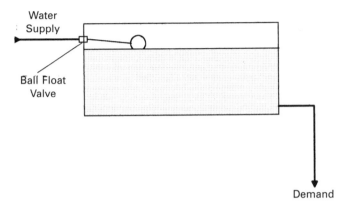

Fig 1.1 Process diagram for break tank.

Some process control applications can be easily put into a category, but some may be a little more difficult, and sometimes may even be a trifle deceptive. For instance, consider the conventional lavatory cistern (W.C.). It is cyclic in its use, but the ball-float valve (ball-cock) is continuously controlling the water level. Although the water level is controlled by the ball-float valve in the cistern, exactly the same design of ball-float valve can be used to constantly control the water level in a break or water header tank (Fig. 1.1) where the outflow from the tank (demand) may be constant, or may vary quite erratically. A typical break/header tank can be found in many houses on the water supply, and the make-up tank of the central heating system, both generally being located in the roof space. The break/header tank application is very definitely continuous, whereas the cistern is apparently cyclic, the cycle being initiated by the action of flushing the toilet. However, should there be a leak in the cistern the ball-float valve will maintain the water level, i.e. continuous control, as in the break/header tank example. As you will no doubt appreciate with these simple everyday examples consideration of what is being done, and how it is done, has already identified a simple control system, and its type. The ball-float valve is a simple continuous control device; it is the lavatory cistern (W.C.) which is a cyclic process. We now see that a cyclic process can often include continuous elements of control. As demonstrated above it is always necessary to identify all the aspects of what is going on to identify the type of control employed. It could be easy to jump to conclusions about the control type, without a little consideration of what is going on.

As you can see from the *loop diagram*, Fig. 1.2, the float valve is a combination of measurement device, controller and final control device built into one unit. The measurement device is the ball-float itself. The final control device is the valve, and the ball crank linkage is the controller. As the water level falls the float follows it. In so doing the linkage, through the bell crank, removes the pressure from the valve, allowing it to open, hence admitting water to the cistern. The reverse action takes place when the water level in the cistern rises, and the valve is closed as the float rises. To compensate for higher water pressures the size of the float or the length of the arm can be increased. If the length of the arm is increased the water level will travel over a greater change to give the same degree of opening of the valve. Hence if the water pressure remains unaltered the length of the arm can be changed to obtain different cut-off levels of the water in the cistern. This is a crude example of changing the proportional action of the control of the valve, i.e. more travel of the **measured variable** (the cistern water level) for the same movement of the final control device (the water valve). This has given us a brief encounter with one of the controller actions to be covered later in this book, along with the development of the other elements covered here. If more information and/or better quality of control are required the individual elements in Fig. 1.2 would be required, so we will now develop those requirements.

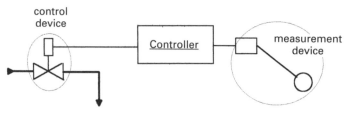

Fig. 1.2 Loop diagram of the ball-float valve.

1.3   Different applications

Examples of various flow diagrams follow, giving a guide as to the type of control most suited for the applications. Although these are from pulp and paper systems they can be likened to many processes for means of identification of which type of control system may best suit the requirements.

The flow diagram of Fig 1.3. shows the arrangement of on/off water valves, for dilution, flushing, etc., plus conveyor, which are operated in sequence to produce batches of pulp for on-going processing. This type of process is easily controlled by either **Programmable Logic Control** (PLC) or **Distributed Control System** (DCS), as are most 'batch' processes. The flow diagram of Fig. 1.4 shows the control items used for consistency control. Due to the continuous nature and speed of the process it is most suited to continuous or DCS control. Figure 1.5 shows a constant level headbox with additional control applied to minimise the overflow/return to stock chest. Due to the nature of the control the loop will have to be fast acting as well as continuous so is definitely only suited to continuous or DCS control.

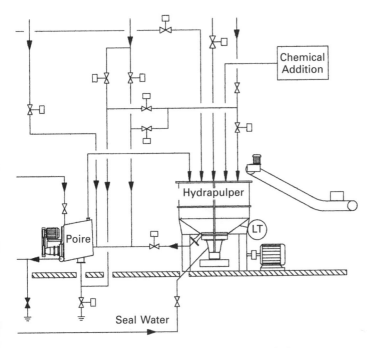

Fig. 1.3 Flow sheet of a typical pulper used in the paper industry.

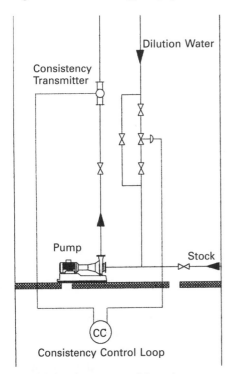

Fig. 1.4 Consistency control for pulp.

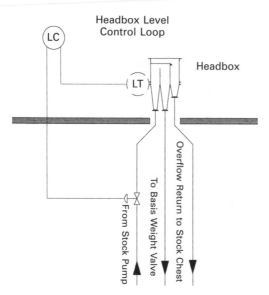

Fig. 1.5 Headbox level control.

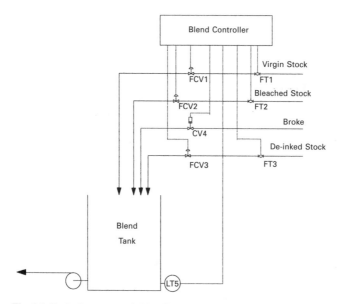

Fig. 1.6 Typical paper stock blending system.

Figure 1.6 shows a typical paper stock blending system where the master control signal is the **blend tank level**. Although there is an element of on/off control for the broke, the remainder is continuous where the control speeds are generally fast (flow control) and hence the most suitable control would be DCS, or a specialised blending control system.

1.4 Operator and management

We should now have some appreciation of the need to break down the elements of each of the control loops, required to make up the overall control of the process, but we must also identify operator and management requirements before completing the process of selecting necessary control equipment to suit the application. Consideration must also be given to the equipment operating environment if a reliable and safe control system is to be applied to the process. As we cover the elements of the system we will take these various points into consideration.

In the case of many simple process control functions little information is required by the operator, or management. Take for example the lavatory cistern again. Providing the level of water is always adequate to flush the toilet, the operator, or management, require no process status information at all. In the case of the break tank, however, there may be a requirement to know the amount of water drawn from the tank, or that the tank has adequate level to meet a demand. The flow information is generally a management requirement, whereas the level information is an operational requirement. In the case of the ball-float valve, to change the level in the tank would require moving the position of the valve up and down the tank side to give the desired level. This is not a practical method of changing the running level (**Set Point**) of the float valve control for applications where the level needs to be changed, so we need to look for a more functional means of controlling the tank level, and provide the necessary information to operators, and management.

Figure 1.7 shows a typical single-loop pneumatic controller for wall or pipe mounting, to provide good basic local control and local display, but without any facility for communications, etc.

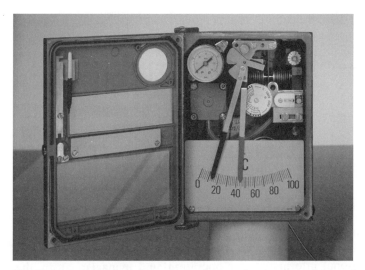

Fig. 1.7 Spirax-Sarco single-loop pneumatic controller (courtesy of Spirax-Sarco Ltd).

A good starting point for identification of operator information is to consider starting and stopping the process, and the information required to carry out these functions, then add to this information all the control functions necessary to run the process continually. Finally we can add to this information such items as raw material consumed by the process (input), and what is produced (output), including all by-products, waste/effluent, etc. There will probably be additional, not so obvious, information about the process required, e.g. electrical consumption, heat input, etc. If the thought procedure applied above, in generating the process flow sheet, is carried out, a large percentage of the required operation functions and information about the process will be identified. This will be especially true when the P & I diagram is generated where items needed for control, along with those for information, are added. Additionally a greater understanding of the process will result which will make the job of applying control that much easier, and potentially more successful. Figure 1.8 shows the example of the break tank with additional instrumentation for operator functions, and management information, described as follows.

- LT   level transmitter
- LC   level controller
- LR   level recorder
- CV   control valve
- FT   flow transmitter
- INT   integrator/totaliser
- FR   flow recorder

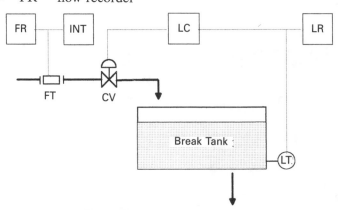

Fig. 1.8 Break tank with additional instrumentation.

The level controller will allow better control of the break tank in all conditions of demand, and will allow change of level set point as required. There is additional information for operation, and management, by the inclusion of the flow transmitter, flow recorder, and integrator/totaliser, plus a level recorder on the level transmitter signal. In many applications there is a requirement for process information if management are to assess running costs, etc.,and historic information to assist in fault finding, production trends, etc. These requirements are satisfied by the addition of integrators/totalisers and recorders as shown in Fig. 1.8.

1.5   Modern trends

The current trend in control is to place a greater emphasis on centralised control/operation. Where a large process is to be controlled, or if there is to be a large addition to an existing plant, consideration should be given to what is already installed, how the plant is to be operated and, if there are to be any changes to the way the plant is operated, to identify whether centralised control could be of benefit. The way a plant is to be operated will significantly influence the type of control equipment and the type of operator interface required. Where large amounts of management data and trend information are required, as well as control of a large number of continuous control loops, consideration should be given to a DCS, since this will save considerable **control panel** costs, as well as installation costs. If the majority of control is of a cyclic nature, a PLC system could be a solution. If, along with operator access to controls, management data plus trend information, etc. is required, a SCADA (**Supervisory Control and Data Acquisition**) system can be added to the PLC. It is therefore essential that as much of the operation, and hence control requirements, is

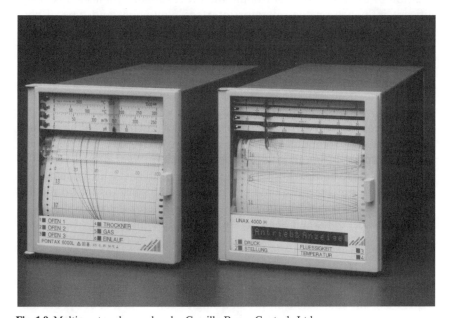

Fig. 1.9 Multipen trend recorders by Camille Bauer Controls Ltd.

identified as early in the project as possible, to ensure that an accurate scope of supply of equipment, and other work needed, is available for cost estimation and planning purposes. SCADA systems can always be added at a later date if required as the PLC system is developed, but if it is envisaged that such an operator interface will be needed in the immediate future it is worth employing such a device when the PLC is commissioned. This will minimise the need for training by ensuring that operators are familiar with the interface before the plant control system becomes quite complex. If the operators are familiar with the SCADA system from the start it will reduce the operator error problems later.

1.6   Special considerations and planning

A good understanding of the process will identify potential pitfalls in the application of controls to the process. The pitfalls are primarily related to the selection of suitable equipment for the process medium and the right type of devices for the task. Record, in detail, any requirements for special process interface materials, waterproof or explosion proof casings, etc., since these can all add substantially to the cost of the project by making instruments, etc. more expensive (more on this in Chapter 2). Careful planning of equipment location can avoid high installation costs or delays. With modern control equipment it is necessary to plan positioning of equipment cubicles at an early stage of a project to maximise utilisation, reduce cabling costs and ensure the right equipment is specified for the job. The location of the operators controlling the process/plant will also have a significant influence on the operator interface equipment and type of control system selected. We will be covering this subject in more detail in the sections devoted to **programmable logic controllers** (PLCs) and **distributed control systems** (DCSs). More important for the choice of control equipment/system are future possible expansions or modifications to the plant, etc. if these are known or under consideration. Purpose-built panels should be avoided if possible, since any modifications required are not normally easy to carry out, and can cause major disruption to the running of plant/process, sometimes resulting in lengthy plant shut-down. Keeping a plant running during times of major modification may give rise to *safety* issues, so choose and plan wisely to avoid the problems which could arise. The way equipment is installed can also influence ease of future modification, etc., so the information to be found in later chapters should be taken into account when deciding on any equipment.

1.7   Existing plant controls

Where existing plant controls are of a single-loop nature, either pneumatic or electronic, encompassing these into either a DCS or PLC system should be considered. This can be done at a later date, but considering it when generating the plant control requirements may give more credibility to the use of a control system rather than continuing with single controllers in dedicated control panels. The actual integration of the existing controls will enable modernisation of the plant, along with giving greater flexibility of control, more operator and management information availability, etc. In the case of control loops where a fast response is required it may be prudent to leave them as single-loop if the control system is to be a networked PLC type. If the controllers have a communications facility, then use this to give the operator information and information for management reporting, etc.

1.8 Summary

We should now have answered some basic questions (in part at least), which gives a developing picture of what will be required to control this process.

1. What type of process are we going to control?
2. Have we got a flow sheet?
3. What type of operation is required?
4. Have we got a P & I diagram?
5. What operator controls are required?
6. Where will the plant be operated from?
7. What information will the operator require?
8. How will this information be presented to the operator?
9. What management information is required?
10. How will the management information be presented?

Hopefully we will have answers to all the above questions, which will give a good 'feel' for the requirements, so the next stage is to start to identify the items of equipment needed to give the functionality required. We will look at each item in the control loop to ensure that the correct information is gathered, and specified, to enable the correct choice of equipment to be made, whether by customer or supplier. If the supplier is identifying the items of control equipment this book should enable the customer to be confident that the proposed equipment will suit the job in hand, since the overall specification should have been generated originally by the customer.

Let us now go back to each individual control loop to gain a greater understanding of the requirements for the items making up the loop. We can consider the basic elements which make up the control loop as follows:

1. measurement device
2. controller

3.   control device/final control element
4.   ancillary devices (recorders, integrator/totalisers, etc.)

Of these items only 1 and 3 will normally be connected to the process medium, and in contact with it, so initially we will investigate these much more fully before embarking on deciding which controller to consider. It is important to remember that unless we employ the correct measurement device and final control device, whatever controller or control system is used will never be able to perform adequately and give good quality control. We can perhaps emphasise this more explicitly for the measurement device using an item of computer jargon, GIGO. This quite simply means: Garbage In gives you Garbage Out!

2 Measurement devices (transmitters)

In the previous chapter we mentioned measurement devices. The success of any control loop is influenced most by the quality of the measurement of the variable to be controlled. Without a very repeatable, reasonably accurate, measurement device, generating the signal for the controller to work on, good quality control will never be achieved. Selection of the measurement device is hence where considerable effort made in the selection process will be well rewarded. In some cases a measurement device is the device to convert the measured parameter into a signal suitable for the controller, but in others it may be the device which converts a measured parameter into another parameter, e.g. an **orifice plate** will generate a differential pressure from the flow of a fluid passing through it, this differential being measured at points upstream and downstream of the orifice plate. A **transmitter** is then required to convert this differential pressure into a signal suitable for the **controller**, **recorder**, etc. For the sake of simplicity we will consider the measurement device and transmitter to be the combination required to generate a signal suitable for control purposes from the process parameter. The type required for the application will depend on the control requirements, as already stated. If we now develop the information needed to identify the unit to suit the application, we can generate a specification sheet which will cover all the process parameters for all measurement devices, and serve as a general specification sheet for all transmitters. Page 15 shows a typical general **transmitter specification sheet**. This sheet can be used to give all the necessary details for the specification of a transmitter of any type adequate for any supplier to provide the correct unit for the service. If the 'manufacturer, model number, and order number' are omitted, the sheet can be used in the 'enquiry document' for accurate pricing, etc. Most measurement device manufacturers and suppliers will confirm that their biggest difficulty in supplying equipment is that the customer regularly supplies inadequate data. Not only does this make additional work for the supplier in establishing the necessary information, but it can often result in the wrong equipment being quoted or even supplied. In the case of the wrong equipment being supplied it may be only an embarrassment, but in some cases it has the potential to be very dangerous. It is therefore essential that all the necessary information is supplied; even if some of the categories do not apply it is better to tell the supplier this by inserting 'not applicable — N/A', rather than letting it be assumed.

2.1   Transmitter specification sheet

We should understand the need for, and importance of, the various sections of the specification sheet so we will cover these in some detail.

Tag number   This is the transmitter identification in the process, and in the control system, and normally defines its type (e.g. PT for pressure transmitter) and sometimes its location. It is unique to the transmitter's location within the process. It is as important to identify any control item's location in the process as it is to identify your own home address.

Transmitter type   This defines the type of prime measurement wanted of the transmitter, e.g. level, pressure, flow, temperature, pH, etc. Although there may be several styles of transmitter to serve a particular parameter the detail of this prime measurement is one of the most important pieces of data on the transmitter specification sheet.

2.2   Process details

As important as the transmitter type, the data should give any supplier an indication as to the style of transmitter required by the process. It is essential that *all* the *process details* are specified to ensure that the transmitter will be suitable for the duty and will operate safely on the process medium. If there are any special conditions of process encountered, *at any time*, these must be included either in the *medium*, or in the *remarks & special details*, e.g. 'Steam cleaning/Sterilisation (specify any chemicals used)'. If the transmitter is likely to be subjected to a large change of medium state this should be included in the *remarks & special details*, e.g. a steam pressure transmitter on a closed vessel which may be subjected to a vacuum condition in certain circumstances.

A word of warning here: *you are responsible for providing equipment to work safely on the process **in all expected conditions**.*

TRANSMITTER SPECIFICATION SHEET

TRANSMITTER TYPE:-   TAG NUMBER:-

PROCESS DETAILS

MEDIUM:-

TEMPERATURE:-

PRESSURE:-

FLOW max.:-   FLOW min:-

ENVIRONMENT DETAILS

HUMIDITY:-
TEMPERATURE:-
OTHER:-
AREA CLASSIFICATION:-

TRANSMITTER DETAILS

PROCESS CONNECTION & SIZE:-

OPEN/CLOSED VESSEL, PIPE etc:-

CALIBRATION MAXIMUM:-

NORMAL WORKING POINT:-

CALIBRATION MINIMUM:-

ZERO SUPPRESSION/ELEVATION:-

SIGNAL CONNECTION:-

SIGNAL STANDARD:-

WETTED PARTS MATERIAL:-

CASE TYPE & SPEC.:-

MANUFACTURER:-

MODEL NUMBER:-

ORDER NUMBER:-

REMARKS & SPECIAL DETAILS:-

PROCESS DETAILS

2.3   Environment details

For long, trouble-free service the conditions of the environment in which the transmitter is to operate must be considered, and specified. Where *hazardous* environments are encountered the applicable code must be detailed (area classification). Even if the environment is just 'external' this must be specified, otherwise the supplier will normally provide the 'lowest cost option', which may not be suitable for the application (e.g. not for external use). Suppliers are in competition and generally each will look closely at their proposed equipment to ensure that the lowest bid is submitted to secure the order. Unfortunately not all will check to ensure that their devices will suit all your criteria exactly before quoting, so examine all quotations carefully to see that the proposed equipment matches your requirements, including the location at which you propose to install it. This piece of information is often overlooked or taken for granted, but it is very important for the durability/reliability of any transmitter. The following pictures show some typical operating environments for control equipment such as transmitters and control valves. When considering the environment it is also essential that you specify any special cleaning to which the devices may be subjected, such as in the food and beverage industries.

The typical large paper mill (Fig. 2.1) shows many areas where mounting of transmitters would be impossible for maintenance access. In many potential mounting places, adjacent to measurement points on pipework, etc., there may be high levels of working temperatures, humidity and vibra-

Fig. 2.1 Typical paper mill environment (courtesy of Fisher-Rosemount).

tion, which will increase maintenance requirements, whilst also making maintenance access difficult.

Fig. 2.2 Typical oil refinery environment (courtesy of Fisher-Rosemount).

Figure 2.2 shows instruments mounted in an exterior environment and hence subject to all weather conditions. In the power plant environment of Fig. 2.3 equipment may be subject to high temperatures and vibration. Fig. 2.4 shows an environment where equipment will be subject to regular, vigorous cleaning routines.

2.4 Transmitter details

Most of the details are self-explanatory and as many as possible should be answered, before putting the sheet out to enquiry, so as to ensure that the right unit is quoted for the application. Some manufacturer specific detail can be left to the prospective supplier to fill in; however, some data is essential.

Fig. 2.3 Typical power plant environment (courtesy of Fisher-Rosemount).

Process connection & size   This must be detailed. Items such as flange size/rating, pipe connection and size, thread type, etc. are required by the supplier.

Open/closed vessel, pipe, etc   Must be detailed since this will dictate if balance connections, etc. are required, and is hence one of the parameters needed to define/confirm the transmitter type.

Calibration and Normal working point   Must be given to ensure that the correct range of instrument is quoted. Any possibility of over-pressure, etc. must be detailed and can be included in the *remarks* section, or as an addition to the working point.

Zero suppression/elevation   This should be stated if known as it may not be able to be accommodated within the rangeability of the proposed trans-

Fig. 2.4 Typical modern brewery environment (courtesy of Fisher-Rosemount).

mitter. Most transmitters can accommodate some suppression/elevation, some more than others. **SMART** types usually have the greatest capability in this area. Some units may require special additions/modifications to accommodate large suppression or elevation of zero.

Signal connection   It is not essential to specify this, but standardising all connections will save installation costs. Avoid special plugs and sockets, non-standard threads, etc., if at all possible.

Signal standard   Must be specified, *including detail of signal isolation.* This data is especially important for PLC and DCS connection — *electronic control systems don't like earth loops!* Many systems now use the 'live zero' to check wiring continuity or damage. For these systems all transmitters must be *fully isolated*, unless *isolated input modules* are fitted to the control system I/O to be used. Additionally some DCS and PLC inputs have electronic protection for over-voltage of analogue inputs, and this will be activated if the transmitter output signal voltage is of too high a value. If you are uncertain as to what specification to use it is always better to consider possible future installations, or moving the transmitter to another location on plant. If full isolation is specified there is every possibility the transmitter can be used anywhere on plant without any signal quality problems arising.

Wetted parts   This can be left to the supplier to quote the most suitable that can be provided to suit the process media details. It may be worth standardising on a specification for all similar transmitters to be of the same **wetted parts**, suited to the most aggressive media to be encountered, so allowing easy interchangeability. For general-purpose applications all wetted parts should at least be stainless steel, or non-ferrous, firstly to stop rusting of the wetted transmitter parts, but also to stop rust staining of the process media.

The remainder of the details on the transmitter specification sheet will generally be supplier specific. Once an order has been placed this information can be added to the sheet so that the transmitter specification sheet can become part of the project documentation, and site records which are essential if any Quality Standards apply to the plant, e.g. BS 5750, ISO 9001.

Examination of what a potential supplier proposes is also a check of whether the transmitter specification has been thoroughly examined by the supplier. A supplier who doesn't comply with the transmitter specification should be treated with caution, or even avoided. Some may not be able to fully meet the specification, and will propose their nearest alternative, stating in the small print that this is the case. Remember it is your responsibility to ensure the transmitter is suitable to operate safely, so ensure all the safety aspects are fully covered.

2.5   Which transmitter?

Having filled in the details of the transmitter specification sheet we will have, by and large, defined the transmitter, but in some cases there are seemingly several types of transmitter to do the same job! So how do we choose which type? The answer is found in the *process details* in relation to the type of measurement required. Certain types of transmitter are more suited to particular applications/industries. In some cases only certain types of transmitter will actually work reliably. Let's take a few common applications.

Temperature measurement

If the process response is slow, and the temperature range is not very high, a **capillary** type instrument might suffice. If the process response is fast, however, the capillary instrument may be too slow to respond, and we should consider a **resistance**, or **thermocouple** type transmitter. N.B. capillaries, and some resistance types, require special facilities for repair, generally necessitating return to supplier, or specialist repairer, which may cause long plant shut-down time, etc. if spare units are not kept on site. Thermocouples, although generally lower in cost, require special compensating cable, matched to the thermocouple type. Installation must ensure that the correct cable core is connected to the correct terminals; otherwise there will be a measurement error of several degrees, caused by secondary junctions. Resistance types are less problematical, but again require all wiring type and connection details to be correctly observed for accurate results.

The thermocouple and resistance elements shown in Fig. 2.5 can be used with local, head mounted transmitters or remote mounted transmitters, for converting the sensor signal to standard instrument signals of 4 to 20 mA, with HART™ in many cases, to give an input suitable for most types of controllers used for temperature control. Figure 2.6 shows a variant of the Spirax-Sarco single-loop pneumatic controller which is fitted with a capillary temperature sensor making the controller suitable for temperature control purposes without the need for any additional temperature sensor.

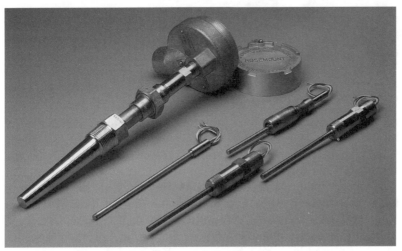

Fig. 2.5 Selection of Fisher-Rosemount temperature sensors, and transmitters (courtesy of Fisher-Rosemount).

Pressure and differential transmitters

The pressure transmitter models probably form the largest proportion of the transmitter types. There are several modes of operation to be found with the **Bourdon tube** types at one end of the scale, and the **strain gauge**, and **solid state** types at the other. The basic **pressure gauge** is normally of the Bourdon tube type and early electronic pressure transmitters were based on the fitting of a potentiometer to the mechanism, giving an output of change of resistance proportional to pressure. This type of transmitter is still available today, needless to say with the advantages that modern technology can provide having been encompassed into the design, and is still a useful transmitter where accuracy is not critical, but low price is the criterion. The main drawback of this type is a fairly high degree of **hysteresis**, and often a tendency to become 'non-linear' as it ages. The Bourdon tube type of unit is also very influenced by vibration, both of the

Fig. 2.6 Spirax-Sarco temperature controller fitted with a capillary filled temperature sensor (courtesy of Spirax-Sarco Ltd).

process medium and general local vibration, and should be avoided if any vibration is likely to be present.

The coming of pneumatic differential pressure transmitters saw the development of the **force balance** design of operation, which was a great improvement on the previous designs, and still represents the majority of pneumatic pressure and differential pressure transmitters today. The major disadvantages with pneumatic transmitters are their requirement for a clean, dry air supply to maintain an accurate output, and their susceptibility to drift caused by local vibration, etc. Force balance uses a positive feedback of the pneumatic output to balance a beam, where both process pressure and output pressure are applied, enabling a 'live zero' to be achieved. This gives a reasonably accurate measurement of pressure and a system which is less influenced by vibration, etc. Although the force balance pneumatic transmitter provides a reliable, reasonably accurate output, it does, however, require regular calibration checks, during which it must be removed from the process. Typical calibration check frequency should be once per year if measurement accuracy is of high importance.

The development of electronic pressure transmitters began with force balance, but other operating systems quickly superseded it, e.g. capacitance, strain gauge, etc., bringing improved accuracy, reduced hysteresis, high reliability, etc. The main pressure/differential transmitters in use by the late 1980s were of the **Capacitance** type, where accuracy is typically $\pm$ 0.25% and hysteresis $\pm$ 0.1% (or less). Other types in common use during this period included strain gauge, which give similar levels of accuracy and hysteresis and are still available today in original and enhanced forms. The

Fig. 2.7 Fisher-Rosemount differential pressure transmitter (courtesy of Fisher-Rosemount).

early 1990s saw the development of more facilities on the transmitter, along with further small accuracy improvements. The development has been in the smart transmitter types, and the communications standards to allow networking of field equipment of *all* measurement and control devices. This development will be on-going for several years to come (up to the year 2000 for the majority), but much of the implementation is planned to ensure that today's equipment will *not* be obsolete tomorrow, which will allow the user to take advantage of future developments as they take place. The smart type facilities are made possible by the use of micro-processors, which means that this development will be applied to all field devices in the fullness of time.

Flow measurement

If we consider fluid flow, such as steam, the most likely choice would be an orifice plate connected to a differential pressure type instrument. This combination has 'turn-down' limitations, typically 3 to 1, if accuracy is to be maintained. The limitation is due to the fact that the differential pressure obeys a square law relative to the fluid velocity. When considering a

Micro-capacitance Silicon Sensor
The heart of the FCX series transmitter is a micro-machined silicon ceramic capacitive type sensor. It is manufactured using plasma etching, photo lithography, glass bonding and other advanced IC manufacturing techniques.
Free from metal fatigue and having adiaphragm deflection of less than 4 micro-meters at full scale it provides excellent linearity with a sensitivity of better than 0.01 percent

Overrange Diaphtagm
Overrange protection to full static pressure rating is given by the overrange protection diaphragm. This diaphragm also reduces the effect of process shock.

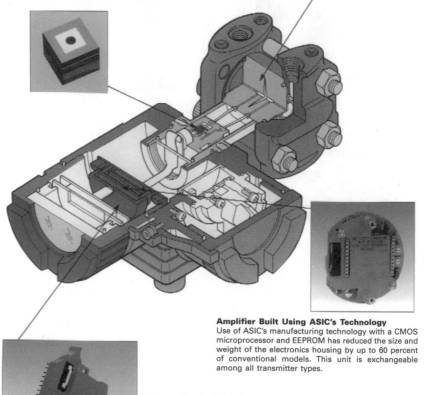

Amplifier Built Using ASIC's Technology
Use of ASIC's manufacturing technology with a CMOS microprocessor and EEPROM has reduced the size and weight of the electronics housing by up to 60 percent of conventional models. This unit is exchangeable among all transmitter types.

Communication Module
The FCX Transmitter is unique. It is designed for both traditional and smart HART[R] type applications. If smart communication is required then a small communication module is plugged into the transmitter. This can be done on-site. It is recommended that the power supply is switched off while plugging the unit in otherwise there is a small risk of EEPROM corruption. Remote communication by a hand held communicator does not interrupt the 4-20 mA signal.

Fig. 2.8 Cut-away drawing showing the internal construction and components of a Fuji FCX-A transmitter (courtesy of Coulton Instrumentation).

Fig. 2.9 Selection of Fuji electric transmitters available from Coulton Instrumentation Ltd.

Fig. 2.10 Fisher-Rosemount 1151 and 3051 transmitters installed on a chemical plant (courtesy of Fisher-Rosemount). The view gives an appreciation of mounting transmitters close together to aid maintenance.

Fig. 2.11 Installation of Fisher-Rosemount differential installed on a carrier ring, plus temperature transmitter to give facility for 'mass flow metering' (courtesy of Fisher-Rosemount).

number of flow measurements using differential pressure transmitters it is worthwhile considering standardising the generated differential by the orifice plate/carrier ring to be the same in all cases, e.g. 100 inches water gauge. This will assist in initial calibration, installation and on-going maintenance. Now alternatives to the orifice plate/differential transmitter combination are available. **Vortex shedding**, and other types of flow-measuring instruments, can offer 'turn-down' of 10 to 1 or more, suitable for use on steam but with a higher cost penalty. The other factor to be considered with these flowmeters is that components are mounted in the pipe. This can cause some pressure loss, generally only small, and they are liable to damage by any inclusions, such as solids, in the fluid. If the fluid is water either of the above units will do the job (with similar limitations), but only if the water is *clean*. If the water is dirty the orifice plate tappings will block, rendering the unit inaccurate or non-functional. An alternative for dirty water, or slurry flow, is the **magnetic flowmeter**. These have a good 'turn-down', are accurate and are a preferred standard in the water and paper industries.

The magnetic flowmeter has the advantage that there is little or no intrusion into the flow stream, which is ideal for media such as paper stock, since there is nothing for the fibre to catch on. But it has a fluid temperature limitation of typically 80°C on most types, although some can be used in temperatures of 120°C with special linings, etc. The magnetic flowmeter should be installed so that the electrodes are horizontal, and the piping should be arranged so that it will remain flooded at all times, even when the flow is zero. Where fibre can catch up, such as in vortex shedding flow transmitters, the caught fibres can develop into blockages, and cause inaccuracy. Other types of flow transmitter available include the **ultrasonic** types. These again do not normally intrude into the flow stream, but need particle matter or bubbles in the fluid to reflect the ultrasonic signal. If the fluid has irregular particle sizes, and/or quantities, 'sensor noise' can result, which may affect accuracy, etc. If there is a large quantity of particles in the fluid many ultrasonic transmitters will give a measurement of the fluid velocity only at the pipe wall, and not averaged across the full diameter of the pipe. Each application must be carefully checked before use of this type of instrument. In many applications of ultrasonic transmitters it is advisable to carry out trial tests for the proposed application. These must prove satisfactory before these transmitters can be classed as suitable for the application. The drawings in Fig. 2.12 show a selection of Krohne magnetic flowmeters for different service applications along with a selection of signal converters which enable the flowmeters to be ranged to suit the required calibration, as well as giving facilities for electrode cleaning, etc.

With all flowmeters the straight pipe lengths before and after the transmitter must be in accordance with the manufacturer's recommendations: typically five pipe diameters upstream and three downstream are the absolute minima; normally upstream should be 10 diameters and downstream five. Generally the longer the straight length of pipe before and after a flow transmitter, the greater will be the accuracy.

Another form of flow metering is the **variable area** meter. The principle is the displacement of a float in a tapered tube as the fluid flow rises through the tube. These are generally used for flow indication, but magnetic floats linked with a position sensor will give a flow signal which can be used for control, etc. The accuracy is not as good as the other methods detailed here, but they do give a low-cost form of metering and are often used for flow switching duties on such applications as cooling water flow, lubricating oil flow, etc.

Mass flow measurement

This measurement is becoming more readily available as the technology has been developed, using the **Correolis** principle. As the name suggests it is a true measurement of mass flow and can be applied to a very wide

... compact design

... separated design

Primary head

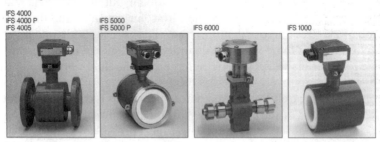

Signal converter

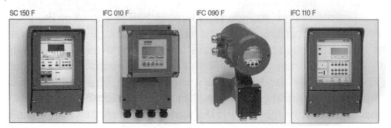

Fig. 2.12 Selection of magnetic flowmeters available from Krohne (courtesy of Krohne Products).

range of fluids, including non-conductive and very viscous fluids. Although expensive, they are generally highly accurate and can give a measurement where previously one was not possible. They are particularly useful where measurement of low flows of viscous fluids is needed, as well as applications where the mass flow of the fluid is required. **Mass flow measurement** using other principles are available, and can give equally good results. Before selecting a particular type it is worthwhile seeking advice as to the suitability of the measurement principle to the process medium, since not all are suited to every application. On applications where small flow rates are achieved using metering pumps the mass flowmeter will often show

Fig. 2.13 Fisher-Rosemount vortex shedding flow transmitter installed in pipe (courtesy of Fisher-Rosemount).

the pulses of the flow derived from the pumping action. In such circumstances it may be necessary to use a measurement input filter (PV damping) on the controller to average the flow and avoid the pulses causing problems with the tuning of the controller.

Level measurement

There are many different types of level transmitters, the most common being the modified **pressure measurement** types. The differential pressure transmitter types are ideal for 'closed' tanks, and the application of diaphragms makes these types ideal for slurries, paper stock, etc. Diaphragms, however, are liable to be easily damaged if the tank contents are severely agitated, especially when including large suspended solids. Other types include capacitance probe, ultrasonic, radar, strain gauge and piezo-electric, etc. The ultrasonic types are ideal for powder and granule level measurement, and for measurement from 'tank top' where there is no process connection available at the bottom of the vessel. **Capacitance probe** types are well suited to conductive liquids, especially those containing debris. Another type which has been about for many years is the **hydrostatic** style level indicators and transmitters. The hydrostatic indicator types are often found on oil tanks and remote locations, since they require no power and are intrinsically safe. There are also several level transmitters made for special applications such as borehole levels, wood chip and grain silo level, etc. which work on operating principles similar to some of those detailed above. Some level measuring devices for difficult media use radioactive source and sensor. These should only be employed if there is no less hazardous type to suit the application. If this type is to be employed a specialist supplier should be used and all recommendations must be fully observed.

Fig. 2.14 VEGA Controls Ultrasonic Level Transmitter (courtesy of VEGA Controls Ltd).

Choice of level transmitter

There are no hard and fast rules governing the choice of **level transmitters** for each application; it is more a case of 'some are much better than others'. In many cases the answer to the question 'what are the site standards?' will define the types and manufacturer to be considered initially. The *environment* and the *process details* will do more to identify the level transmitter type to choose. Generally if the medium is of a slurry, small

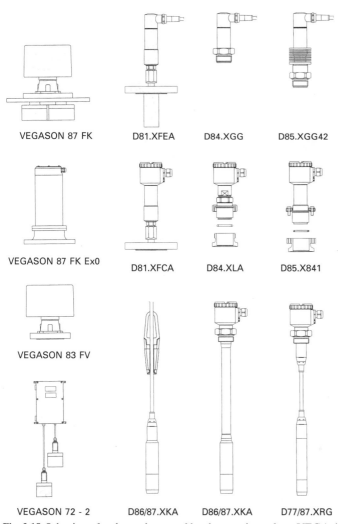

VEGASON 87 FK   D81.XFEA   D84.XGG   D85.XGG42

VEGASON 87 FK Ex0   D81.XFCA   D84.XLA   D85.X841

VEGASON 83 FV

VEGASON 72 - 2   D86/87.XKA   D86/87.XKA   D77/87.XRG

Fig. 2.15 Selection of styles and types of level transmitters from VEGA (courtesy of VEGA Controls Ltd).

granules or fibre a **diaphragm** or ultrasonic type is the better option. If the vessel is closed a differential pressure type is required, and for effluent sumps, etc. a **probe** type is possibly more suited. If you are not familiar with 'what to put where', send your enquiry to several leading general transmitter suppliers. The result from each should be similar, providing the details in the transmitter specification sheet are complete and correct. It is obviously worth taking into account standardising on types, ranges and manufacturer to gain the benefits of interchangeability, etc. as well as considering the cost advantages of a package of supply of controls and transmitters. If there is a possibility of any water vapour being present above the surface of the medium, the ultrasonic types may not always measure

Fig. 2.16 Fisher-Rosemount 1151 Diaphragm Level Transmitter (courtesy of Fisher-Rosemount).

the level correctly, due to scatter of the ultrasonic beam off the water vapour. This situation, of potential water vapour on the surface, should be detailed on the transmitter specification sheet to ensure the prospective supplier can be aware of a possible problem. Figures 2.14–2.16 show different types of level transmitters.

Speed measurement

Until recent years the usual device used has been a **tachometer** generating a dc voltage which is directly proportional to the speed of rotation. Now that **encoders** are more readily available these are being used more frequently for high accuracy, and have the advantage of pulse/frequency output, which is required where digital input to the controller is needed. These devices save the inaccuracy of analogue to digital signal conversion, and as the units do not employ brushgear they tend to be more reliable and require less maintenance.

2.6   Special service measurements and transmitters

The transmitters detailed above will cover most of the day-to-day applications encountered, but there are some services where a purpose-designed transmitter will be needed, e.g. stock and slurry consistency, oxygen, high temperature humidity, pH, conductivity, etc. With some of these there is still sometimes a choice of types.

Fig. 2.17 DeZurik Rotary Electronic Consistency Transmitter.

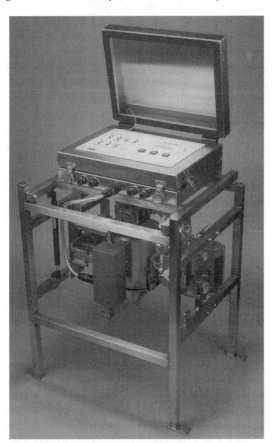

Fig. 2.18 Croft Instruments Ashcon Optical Consistency Transmitter.

Consistency transmitters

These measure % suspension of solids/fibre in water, by weight and are found mainly in the paper industry and sometimes in cement, sewage and similar applications. There are two main types, **shear measurement**, and **optical**, plus a few lesser developed techniques. The shear measurement are either measurement of torque on a rotary element or deflection of a blade. In paper stock applications the range is generally limited to between 1 and 6% consistency, although with some units up to 10% can be measured on some types of stock. The installation of these units is critical for good results. The manufacturers give very precise installation data, which should always be observed. The main problem with these units is that stock flow must normally be maintained within very precise limits to ensure an accurate measurement, and the scale of calibration should be about 2% anywhere within the range limit (e.g. 3 to 5%, or 2.5 to 4.5%) for the best results. With the application of microprocessors the rotary shear measurement types are less affected by flow variation on many stocks, but in some situations flow influence will still be present. With optical principles the range can be from clean water up to 5% consistency, but again with a 2% calibrated scale. This type is less affected by flow variations, but flow must always be above a base limit. They are, however, affected by the mix of fibre and solids (ash), some types now giving outputs of consistency and solids. With optical units occasional cleaning is necessary when used on many services. For any consistency application it is worthwhile carrying out a trial if there is no in-house experience of the type under consideration. One of several types of consistency measurement transmitters for pulp and paper stocks, as well as cement and sewage slurries, etc. is shown in Fig 2.17. Figure 2.18 shows another type of consistency measurement transmitter able to measure very low concentrations of solids. It employs a microprocessor and is capable of giving more information than consistency, or suspended solids, but the prime measurement is the one which should be employed for control purposes, the remainder being for error correction and information.

Suspended solids and turbidity meters

These are generally very similar in their operation to optical consistency transmitters.

Oxygen transmitters for gases

There are again several types to choose from, but the most popular today are the **Zirconia** probe types. The Zirconia characteristics allow a direct measurement of the oxygen present in the gas without need for samples to be cooled, etc. and are hence ideal for oxygen measurement in furnace, and boiler flue gases. Like the consistency transmitters the installation is critical for good results. A recent development of the Zirconia type has been the **high temperature humidity** transmitter, giving a lower cost alternative to the very expensive **infra-red** types for this application.

pH transmitters

The units available all work on the same principle of measurement, with individual refinements. Very small changes in the electrical characteristics of the fluid are measured with reference to a known fluid, normally a salt solution. pH transmitters have to be periodically 'buffered' with solutions of known pH to establish the calibration. Most pH transmitters use glass electrodes and are hence very delicate. Installation is of prime importance for good results, remembering that access will be regularly required for maintenance, buffering, etc.

Conductivity transmitters

These can be put in the same category as the pH transmitters above, but do differ in type and maintenance requirements. The need for regular calibration checks is far less, although they are still required. Generally no reference solution or buffer solutions are required. Many microprocessor types need a minimum amount of calibration checking.

Oxygen transmitters for liquids (dissolved oxygen transmitters)

Generally these are very similar to conductivity transmitters in their operating principle.

Analytical transmitters

The three types above, oxygen, pH, and conductivity transmitters are generally grouped under **analytical transmitters**, along with other special analytical types of units.

Others

Other types of transmitters are available for different duties, but will require specialist knowledge for the application in hand. One such specialist field is **moisture and basis weight measurement** used on paper and plastic film production. These units are sometimes referred to as **gauging systems** and are very expensive. When such measurements are required the relevant suppliers should be invited to look at the application. They will advise on all aspects of the equipment needed and the installation requirements. In addition they will give an overview of the service and support they can give to the equipment and, if required, financial justifications (e.g. estimated cost savings/payback). As you will appreciate, such specialist units are a costly investment; any choice should include the production team, etc. to ensure that all the users will be comfortable with the equipment to be supplied. Although costly the product brings a payback in quality improvement which justifies the investment. Equipment of this nature is generally of several weeks/months delivery, so early identification of requirement will help the project planning. As hinted above, this type of measurement needs a considerable degree of maintenance to ensure that the operation and calibration are always kept to the highest standard.

Use of radioactive sources

There are some transmitters which utilise **radioactive** sources in the measurement technology, e.g. **basis weight measurement**. If these are to be considered ensure that all the applicable regulations, etc. are observed before proceeding with installation. With most of these units specialised maintenance facilities will be necessary, and a specified member of staff responsible for **isotope handling** (RPS) may need to be available on site.

HART™ communication protocol

Smart control equipment (field devices) have various types of protocol employed to enable communication to be made with the field device from a remote location. One such protocol, which is becoming an industry standard, is HART™ **(Highway Addressable Remote Transducer)**. The transmitter is 'hard wired' in the normal way, carrying an analogue, 4 to 20 mA, signal. The communication is superimposed on the signal wiring, and is used to interrogate the transmitter for calibration data, etc. and can be used to change calibration to whatever is required, without the need for disconnection or removal to a workshop of the field device. The act of communicating with the field device in no way affects the analogue, 4 to 20 mA signal. HART™ is now being developed for multi-drop **Fieldbus** networked field devices and, although developed by Rosemount, is available for use by any manufacturer and can be applied to most types of electronic measurement device. The device shown in Fig. 2.19 is used for calibrating SMART transmitters which utilise the HART™ communication protocol. It can also be used for checking HART™ transmitter outputs, etc.

With all types of transmitters try to standardise on types, manufacturers, ranges, etc. if at all possible. This will cut down on spares, and simplify installation. It is also worthwhile standardising on calibration of transmitters if possible, as mentioned above with respect to orifice plate differentials. With modern electronic and digital transmitters the accuracy is very high, stable and repeatable, and in many applications standardising of some transmitter capsule ranges will not give rise to noticeable errors. A simplified scope of supply can also give a better project discount in some cases.

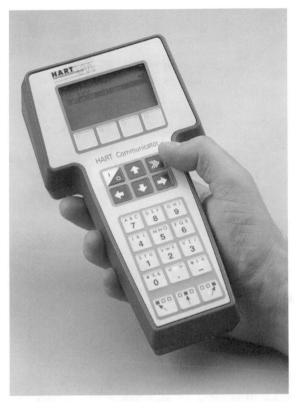

Fig. 2.19 HART™ hand-held communication/calibration unit (courtesy of Fisher-Rosemount).

2.7   Installation of transmitters

With all transmitters the way they are installed is very important for long-term reliability and quality of measurement signal. A transmitter which is inaccessible will be the one to give trouble (one facet of 'sods' law'). Manufacturers' installation data will give all the necessary information you need for a good, reliable unit, but there are a few basic considerations which can help at the planning, and design stage.

1. Do not subject transmitters to excessive vibration, unless this is the parameter to be measured.
2. Give adequate support to heavy transmitters and associated parts including the cables, etc.
3. Mount transmitters where they will measure what you require! This sounds cock-eyed but experience has often shown control instability resulting from this, e.g. a pressure transmitter mounted so close to a pump outlet it measures the pump pressure pulses more than the process pressure, or mounted adjacent to the control valve giving

Fig. 2.20 Technician using HART™ hand-held communicator to check the calibration of a transmitter (courtesy of Fisher-Rosemount).

unrequired **derivative** control action, or uncontrollable control loop oscillation.

4. The right measurement range for the duty. If you want control at 100 psig to within ± 1 psi, a transmitter calibrated 0 to 1000 psig will give poor results! Generally allow a minimum of 20% above and below the maximum and minimum desired operating point for the calibrated range.

5. Avoid capillary type level and differential pressure transmitters where large environment temperature changes may be experienced. The capillary can act as a secondary 'temperature transmitter' which can cause errors in the measurement.

6. Generally avoid mounting any transmitter in high temperature environments.

7. Locate instruments together if possible; this can save installation costs and make maintenance easier.

8. Avoid mounting electronic instruments adjacent to big electrical

power devices and power cables.
9. Ensure pneumatic devices have a clean, dry, air supply, regulated to the correct pressure.
10. Avoid locating instruments where they make good steps or handles!
11. If the measured medium is liable to freeze the transmitter should be housed in a heated, purpose-made housing and, if applicable, the impulse piping lagged and heat traced. Anti-freeze should be avoided as this may contaminate the process medium being measured and 'work' out of the impulse pipework.
12. Observe all regulations and guidelines, especially for installation in hazardous areas.

The above may seem a little basic, and obvious, but if any of the above points is not considered problems will result sooner or later. Let us now consider installation of transmitters in more detail.

2.8   Process connection to transmitters

The following installation recommendations will hopefully give the reader a guide to achieving a good working installation for most general applications. They will form a good basic instruction to an installation contractor and ensure a good measurement of the process variable. These recommendations are not proposed to replace manufacturers' recommendations, but to supplement them. *Please check with your safety representative to ensure all safety requirements are met.*

GENERAL

1. All pipework should be cleaned of burs, swarf, jointing compound, etc. before final connection to instruments.
2. All gaskets to be of correct size and material for the duty, etc. and ensure gaskets do not foul transmitter diaphragms, etc.
3. All nuts and bolts to be fitted with washers. In wet environment use non-ferrous nuts and bolts, washers, etc. Also external mounted transmitters should be fitted with non-ferrous fixings.
4. Observe all process flow direction arrows on transmitters.
5. Do not cut or kink capillaries to assist installation.
6. Remove packing material at time of installation. (Plastic bungs do not assist measurement!)
7. Do not fit electronic instruments on vessels, etc. until all welding is completed.

Fig. 2.21 Fisher-Rosemount transmitters installed on a refinery (courtesy of Fisher-Rosemount).

PIPED CONNECTION TO WET PROCESS (low temperature and pressures)

1. If possible always provide isolation between process and instrument (transmitter). If connection and isolation to the process is not easily accessible fit additional isolation at the transmitter to aid maintenance.
2. Arrange pipework to rise or fall from process connection avoiding air locks; typically a minimum fall of 1 in 20 will suffice on many applications.
3. On falling pipework connections 'tee' instrument off pipework and form a valved, plugged, vertical dirt trap and drain.
4. Use a jointing compound if there is a possibility that pieces of 'thread tape' can contaminate process or foul transmitter workings.
5. Where ferrous materials are cut, including pipe threads, paint exposed cut areas.
6. Ensure threads on pipework match instrument connections. Avoid

over-tightening fittings into instruments to achieve a leakproof connection!

7. Support all connection pipework adequately.
8. If the process medium is hot allow adequate length of impulse pipework to ensure the working temperature of the transmitter is well within specification, *Winter and Summer*.

FLANGED CONNECTION TO PROCESS (low temperature and pressure)

1. 1, 4 and 5 above will apply.
2. Mount instrument according to manufacturers' instructions with respect to orientation, etc.
3. Ensure gaskets are of correct material and size and do not foul diaphragms, etc. especially when compressed.
4. Where remote seals are used do not strain capillaries and support them properly, especially excess lengths which should be coiled carefully and supported without twisting.

TEMPERATURE TRANSMITTERS

1. Observe 4 above for capillary devices.
2. Use a thermowell or pocket if at all possible. If no pocket is fitted put a warning on the connection to identify that no thermowell or pocket is fitted.
3. Do not use 'thread tapes' if this will possibly contaminate the process.
4. Avoid joints in wiring between sensor and transmitter. Keep this wiring as short as possible to avoid electromagnetic interference on the signal.
5. Run wiring between sensor and transmitter well supported on tray or in conduit or trunking.
6. When installing thermocouple-type instruments ensure the correct compensating cable is used, and observe the polarity of the cores when connecting to both thermocouple and instrument. If junction blocks are necessary ensure they are of the correct type and fitted observing polarities, otherwise secondary junctions will be generated which will give large errors to the measured parameter.

PROBE-TYPE TRANSMITTERS IN TANKS

1. If agitation is present in the tank ensure that the probe is adequately supported in the tank and positioned in a sheltered zone if at all possible.
2. Avoid submerged joints. If joints are necessary ensure that these are

of the correct type recommended by the instrument manufacturer.
3. Ensure that mounting will allow removal without draining tank, etc. if at all possible.

CONNECTION TO DRY PROCESS

All the above will generally apply but special consideration should be given to the medium. Where gases are involved double isolation may be considered where the final isolation point is 'lockable'. In the case of powders use close fitting measurement devices to avoid impulse lines blocking.

CONNECTION TO STEAM SERVICE

1. Ensure that all materials, and fittings are suitable for the pressure and temperature of the medium. N.B. On flange ratings, pressure ratings fall as temperature rises.
2. Ensure adequate allowance is made for expansion and contraction in all connection pipework and mountings. *Transmitters must not be subjected to expansion forces.*
3. Single or double isolation must be provided at the point of connection to the process and, if necessary, at the transmitter. The latter can save draining pipework every time the transmitter is removed.
4. Where possible connect with falling pipework to transmitter (1 in 20 minimum, vertical if possible), with transmitter 'teed' off the pipework so a dirt trap and valved, plugged drain is formed. It can be advantageous to incorporate a condensate pot at the process connection point to overcome process noise generated by flashing condensate at the connection point, especially on connection to an orifice plate or carrier ring. In the case of the orifice plate or carrier ring, both pots must be on *identical horizontal level of process connection* at the pots.
5. Connection pipework must be adequately supported (e.g. horizontal at 1 m pitch, vertically at 2 m pitch).
6. On low range measurements pipe fall should be no less than 1 in 4 (vertical if possible).
7. Connection pipe bore should be no less than 10 mm (3/8th inch) and in general 15 mm (1/2 inch) should be employed, and should present a minimum of 1 m condensate head between process connection and transmitter, or more if the temperature presented to the instrument would be above 60°C in any operating condition (or above manufacturers' recommendation).
8. *N.B. All connection pipework forms part of the 'pressure system' and as such is subject to the applicable regulations.*

CONNECTION TO VACUUM SERVICE

If the considerations applied to *steam service* above are applied to vacuum, a satisfactory installation will result.

HIGH PRESSURE and TEMPERATURE CONNECTIONS

All the aspects of *connection to steam service* above will apply with the addition of the material and fittings ratings for the duty. The procedures for installation and maintenance, however, will be in accordance with the applicable regulations.

In most applications industry standards already apply. These should be observed for all installations. Please check with your safety officer/manager to ensure all regulations, guidelines and safety requirements are met.

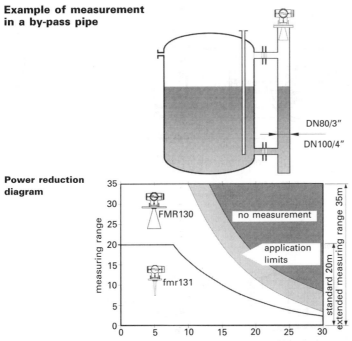

Example of measurement in a by-pass pipe

DN80/3″

DN100/4″

Power reduction diagram

Fig. 2.22 Arrangement of Application of Endress + Hauser micropilot FMR 130 microwave Level Transmitter in a by-pass pipe on a high pressure vessel to avoid spurious readings generated by surface movement of the liquid (courtesy of Endress + Hauser).

2.9 How NOT to install transmitters

There are many publications giving details of how to install measurement devices and all manufacturers should now provide full details of how to install their own equipment. Some examples are shown in Figs. 2.22–2.25, courtesy of Endress + Hauser, for some of their range of transmitters.

Installation instructions pressure sensors

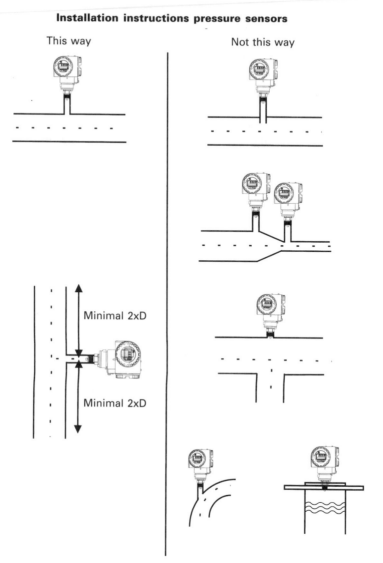

Fig. 2.23 Installation details for Endress + Hauser Cerebar Pressure Transmitters (courtesy of Endress + Hauser).

Unfortunately there are not many publications which give detail of how transmitters should not be installed and why. Most installation details are given with respect to measurement accuracy and reliability, not necessarily potential control problems. There follows a brief outline of how not to install some transmitters, designed to allow you to identify a correct installation when you see it and to know why some installations are incorrect. The loop diagrams, Figs. 2.26 and 2.27, give some examples of 'How not to install transmitters'.

The pressure in the vessel of Fig. 2.26 is the process parameter which

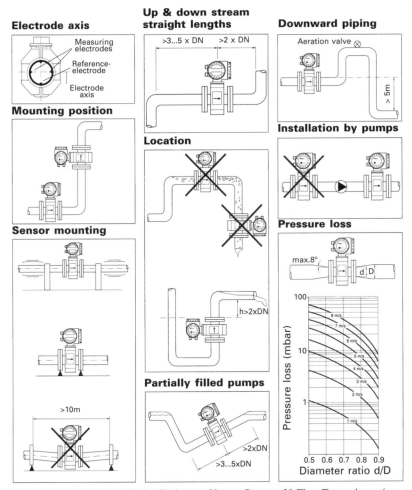

Fig. 2.24 Installation details for Endress + Hauser Promag 30 Flow Transmitters (courtesy of Endress + Hauser).

must be controlled. The correct position for the pressure transmitter is therefore in position 2 on the vessel. If, however, position 1 is chosen, the transmitter will see the pressure in the pipe as the prime measurement, feeding this to the pressure controller (PC) immediately. This will cause the pressure control valve (PCV) to be modulated to regain the pressure set point, which is the pressure in the pipe! This causes the pressure in the vessel to lag significantly behind, and possibly never attain, the required value. To overcome this problem it would be necessary to run the pressure control with a high off-set to the pressure control set point. Additionally the closer the location of the transmitter to the control valve (PCV), the more unstable the control will be, since every movement of the valve will have rapid effect on the measured pressure, feeding this change to the controller (PC) immediately. If, due to mechanical

Installation guidelines

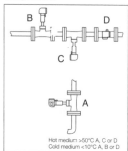

Mounting and medium temperature

Position of pressure (P) and temperature (T)measurement

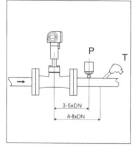

Mounting in insulated piping

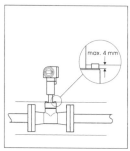

Fig. 2.25 Installation details for Endress + Hauser Prowirl 70 Flow Transmitters (courtesy of Endress + Hauser).

constraints, the pressure transmitter (PT) cannot be located on the vessel (e.g. the vessel is a rotating cylinder), the pressure transmitter (PT) should be located as close to the vessel as possible, and the control valve (PCV) at least 20 pipe diameters upstream of the transmitter (PT). This arrangement will give a reasonable time lag in the control system, enabling a more stable control situation to exist. If safety valves are to be fitted, ensure these are located after the pressure transmitter and definitely not between the transmitter and the control valve. If the safety valve is installed between the transmitter and the control valve, and the running pressure set point is close to the safety valve lift setting, the safety will lift if the opening of the control valve causes the pressure to rise above safety set

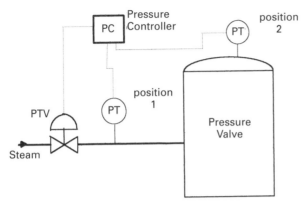

PT in position 1 can give instability and errors
PT in position 2 is ideal

Fig. 2.26 Control of pressure in a vessel.

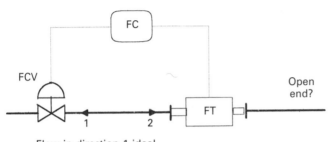

Flow in direction 1 ideal
Flow in direction 2 possible instability

Fig. 2.27 Control of flow.

pressure. The transmitter will see not the rise in pressure but the fall resulting from the safety lifting, and hence open the control valve more in an effort to establish pressure to set point. This is a typical case of interaction between two control devices due to bad design, which could easily be avoided. The safety valve should be installed to protect the vessel, not the length of pipe between the control valve and the transmitter! Not only will the lifting of the safety waste steam, but the control of the pressure will be bad as well.

In Fig. 2.27 the control valve (FCV), should be located at least 10 pipe diameters away from the flowmeter (FT), no matter what the flow direction. Flow in direction 1 will ensure that the FT is not subjected to wide pressure variations as the FCV modulates, which could cause metering errors if the transmitter is of an orifice plate/differential pressure transmitter combination, in comparison to the arrangement in flow direction 2. If the flow in direction 2 passes through the FT and then the pipe is open-ended, e.g. into an open tank, the flow meter bore can easily become empty, with the result that metering errors (zero drift) will ensue. Arranging the pipework to ensure the FT remains 'flooded' in this situation will avoid the tube becoming empty, but if there is insufficient back pressure instability can still occur.

With any transmitter and associated control valve (control device), ensure that the transmitter is a true measure of the process parameter required and that the distance between the two devices will adequately filter the disturbance on the process parameter generated by modulation of the control device. The response time to any such disturbance should be measurable in seconds (or possibly fractions of a second), not milliseconds. Remember this response time will change as process load changes, and it may also change with change of set point. Any measurement device installed too close to any control device will pick up 'noise' from the control device, which may possibly cause severe control problems, such as instability and, at worst, inability to control at all.

3 Control devices (final control element)

We have detailed how we measure the process parameter we wish to control and have also briefly covered cyclic and continuous control. The control device (final control element) has also been touched on and in this chapter we will cover control devices in much more detail. For good control all the components making up the control loop, and how they are installed, are of great importance in achieving our goal. Unfortunately, in any control loop, the control device is often the component where least trouble is taken to ensure that it is of the correct type, suitability and size for the job in hand. Primarily this is most likely to be due to the fact that the control device is quite often selected by the process design engineer, who may not be aware of the ideal type of control device for the application or of the need for accurate sizing of control valves, etc. Many times over- or under-sized control devices will cause alarming control problems, including instability or even inability for the loop to be controlled. We have already stated that for ideal control the field devices (measurement and control devices) should run, for a normal control set point, at 50 to 60% of their calibrated range, which includes the control device. Bearing this in mind, if the control device is normally working almost at its maximum, or minimum, it is probably sized incorrectly. The control device is also the item in the control loop which normally does most work and hence suffers the highest degree of wear and tear. Another major problem of control is that often the control device receives little maintenance, when possibly it requires more than any other component in the loop, purely by the nature of how it operates. Often the control device is placed in an inaccessible location, a problem which could possibly be avoided if identified during the process plant design stage. In our selection process for the most suitable control device, the factors to consider must include matching its durability to the duty it must perform, along with gauging the level of maintenance required to keep it in an acceptable operating condition, as well as the sizing of it. Let us now identify some of the more common control devices:

1. **control valves**
2. **variable speed drives** (pumps, etc.)
3. **directly coupled actuators** (found on burners, etc.)
4. others

3.1   Control valves

Control valves possibly make up the most common of all control devices. They have been around for many years and are available in many different designs and forms. Although they have been about for a long time this doesn't mean to say they are old fashioned. The design of the control valve is still being developed, taking advantage of modern materials, computer simulation and other technological improvements. Still very common today are the pneumatically actuated types. Although modern control uses the current technology of electronics and microprocessors, compressed air is still the most useful medium for the power source in actuation of modulating control valves. This power source along with electronic and microprocessor positioners and actuators means the control valve is right up-to-date. Other power sources are used for actuation other than pneumatic, e.g. electrical, hydraulic, gas, etc., but to date the pneumatic actuated control valve is still very cost-effective, reliable and easy to maintain. As with any pneumatic device clean air is essential for reliable operation, and every control valve should always be fitted with an air filter (25 micron or better) on the air supply to it.

When considering control valve suppliers there are many to choose from, some able to supply a complete range of types and others specialising in one or two types only. There are also many valve agents who cover a number of manufacturers and are hence able to offer a control valve for most applications. The selection of the correct type for the application is not always a simple task! The process will quite often give an indication as to which type of control valve to use, but not as easily as in the case of a measurement device. We can split the control valve into two parts for consideration when embarking on the selection process. The **body** is the part where the control is effected on the process medium, and the **actuator** is the drive mechanism which converts the control signal into a force on the body's moving component to bring the effect about. The first stage of the selection is to identify the body type required. The pressures and temperatures to be encountered will be the main consideration once the process medium itself has been defined. Because there are so many factors to be taken into account the generation of another specification sheet should be considered. A general **control valve specification sheet** will be found on the next page; it is broken down into the necessary blocks of

CONTROL VALVE SPECIFICATION SHEET

CONTROL VALVE TYPE:-   TAG NUMBER:-
PROCESS DETAILS

MEDIUM:-   FLOW RATE MIN:-

S.G.:-   FLOW RATE NORM:-

VISCOSITY:-   FLOW RATE MAX:-

TEMPERATURE:-   PRESSURE:-

ENVIRONMENT & PIPEWORK DETAILS
HUMIDITY:-   TEMPERATURE:-

OTHER:-   PIPE SIZE:-
(HAZARDOUS Area Class Etc.)
PIPE MATERIAL:-   ORIENTATION:-

VALVE BODY DETAILS

PROCESS CONNECTION & SIZE (RATING):-

D.P. CLOSED:-   D.P. @ MIN FLOW:-

D.P. @ NORM FLOW:-   D.P. @ MAX FLOW:-

BODY STYLE:-   BODY MATERIAL:-

INTERNALS:-   SPECIAL REQUIREMENTS:-

ACTUATOR

TYPE:-   POSITIONER:-

ACTION:-   AIR FAIL ACTION:-

CONDITIONING/CAM:-   SIGNAL INPUT:-

SIGNAL CONVERTER:-
REMARKS & SPECIAL DETAILS:-

ORDER NUMBER:-

MANUFACTURER:-
MODEL NUMBER:-

Fig. 3.1 DeZurik-type Maxam RCV control valve with diaphragm actuator and positioner.

data required for a control valve supplier to identify the type of unit required. We have covered some of the data which is similar in the transmitter specification sheet, but some data has to be expanded to allow the correct sizing, as well as the type, to be achieved. Let us look at some of the data in more detail and find out why it is required.

Process, environment and valve body details

If the environment is of a hazardous nature full details must be specified on all enquiry documentation.

Medium   The type of fluid will influence the body and internals materials to ensure the control valve will not fail due to rapid corrosion, melting components, etc. The style of body will also be influenced by the *medium*, e.g. blockages, high wear due to highly abrasive medium, etc.

Temperature and pressure   These parameters will affect the way the controlling surfaces will wear when working, as well as defining the body strength, suitable materials, connection sizes, etc. These parameters will

Fig. 3.2 Siemens control valve with actuator and positioner.

also influence the thrust required from the actuator to keep the working parts of the valve in the required position under working conditions and hence will be used to size the actuator. The pressure drop across the control valve is possibly the most important factor of all. This will determine the valve size when related to the required flow, but can also generate the most wear within the valve body.

Flow rates To determine the size of a control valve for any application the Cv is required. The Cv is the valve sizing, or flow, coefficient and is defined as 'the number of U.S. gallons per minute of water at 60°F that will flow through a valve with a pressure drop of 1 pound per square inch'. To establish the correct body size, the Cv at minimum, normal and maximum flows is necessary. The flow rates combined with the pressure drops (DP) at the various flows are used to calculate the valve Cv at each flow, and hence the size of valve required. The Cvs for different types of valve are experimentally determined by the valve manufacturers and the 'basic liquid sizing equation' is used:

$$Q = Cv \sqrt{\frac{DP}{G}}$$

where Q is the liquid flow rate, G is the specific gravity of the fluid and DP the pressure drop across the valve.

We do not need to get into the detail of how the Cv for each valve is actually obtained, but we do need to know what Cv is required of the selected valve for the application. Control valve manufacturers and suppliers can provide a sizing service for any valve/application, so if the data is provided correctly the sizing can be done by the supplier. Many suppliers now provide software packages for the sizing of their valves, but these should be used by the project engineer only if many valves are required, and even in these circumstances it is better to allow the supplier to do the final sizing, for guarantee and safety reasons, as well as a back check on your sizing! If you use such software it is essential that the valve supplier gives you a demonstration of how you should use it for all the different types of calculations and types of valves the software can perform. The project in hand may not require all the calculations initially, but you will find them very useful later when you may be asked to give details of what the potential maximum flow through a section of plant is. It will also be useful when you find that the process design has changed and you need to ascertain if the selected valves will still be suitable. Finally for some applications, e.g. safety valve sizing, you will need to know the maximum flow that may be encountered if a control valve fails full open. Use the software to give an indication, but get the control valve supplier to confirm the figure.

Types of control valve

Butterfly   Sometimes known as a disc valve. It takes the form of a disc normally mounted central on a shaft. By rotation of the shaft the disc will travel from a position closing off the flow when across the pipe, to fully open at 90° to the pipe, when allowing the maximum possible opening and flow. 'Offset disc' butterfly valves are available for special control applications; they tend to give a more linear characteristic but are also slightly more expensive. By use of different seat materials they can be made to suit water, gas and steam control.

Globe   Also known as **plug and seat**-type valve. The body is arranged to give a fluid flow up through an orifice. A plug lowered into the orifice will

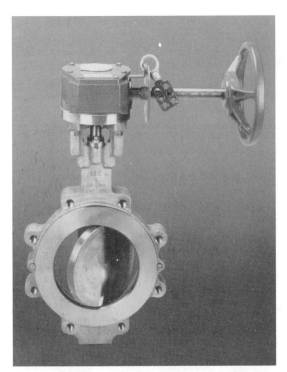

Fig. 3.3 DeZurik BHP high performance butterfly valve.

Fig. 3.4 Spirax-Sarco pneumatically actuated globe body control valve c/w positioner.

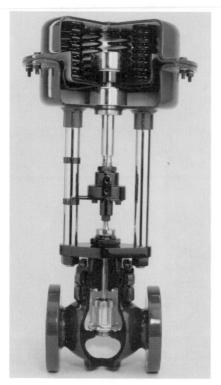

Fig. 3.5 Cut away detail of Spirax-Sarco control valve (courtesy of Spirax-Sarco Ltd).

give control of the flow from maximum to zero when the plug is fully inserted into the orifice or seat. Normally the seat material is metal.

Ball   By inserting a ball with a hole through it in line with the flow direction in the valve body, a control of the flow can be achieved by rotating the ball through 90°, similar to the butterfly valve.

Profile or segmented ball   By the addition of a profiled seat or shape to the orifice in the ball segment, a more accurate control of the fluid flow can be achieved. If the profile is in the form of a 'vee' the flow will be directly proportional to the area of opening, so to make the relationship of angle of opening directly proportional to open area will give a linear valve flow response. Other responses can hence be achieved by profiling the seat or ball segment.

Pinch   Generally a rubber lined valve which effects throttling of flow by applying pressure between the body and the rubber lining. This closes the lining across the pipe and hence restricts the flow. It is a very crude valve,

Fig. 3.6 DeZurik VPB vee-port valve with Powerac actuator and positioner.

is not very repeatable, and hence not good for control purposes unless no alternative is suited to the medium.

Spade or gate As the name suggests these are very basic control valves employing a sliding spade across the pipe to effect control of flow. Generally these give very poor flow control and should generally be employed only for diversion of flow and not control of flow. This type of control valve should definitely not be used for control of pressure. On pressure type applications it should be used only for isolation, and then only the 'parallel slide' type specially designed for this purpose.

Control valve body

The completed control valve specification sheet data will enable a supplier to size and select the valve body to suit the application, but for us to feel comfortable with the various possible proposed types let us now look at which type and style of valve body suits which service.

Water type fluids   For fluids such as water we can use several types of body. The lowest cost type is typically butterfly, whereas the highest cost type is possibly profiled ball. Butterfly types are well suited to general applications where there is little inclusion of fibre, grit, etc. in the fluid and the quality of control is not critical, e.g. tank level, etc. This type of valve is suitable for flow type control, will not normally give tight shut-off, unless 'soft seated', but is not desirable for pressure control. The opening to flow characteristics are generally non-linear giving a rapid rise at small openings and little throttling at large openings. They are, however, low-cost and if their characteristics can be tolerated, are an economical valve to use. Special styles of butterfly valves are available where the disc is profiled, or off-set, to give a more linear characteristic. Unfortunately as these styles make up a minority group within the butterfly valve type the cost increase may not be worthwhile considering when a different body type may be more suitable for the application.

Fig. 3.7 View of pulp mill showing instruments and control valves (at high level). Access to the control valves in this view is obviously not easy! (Courtesy of Fisher-Rosemount).

For better/more accurate control the globe-style body is a mid-range price option. The globe or plug and seat style gives tight shut-off on a hard seat and a better characteristic of 'opening to flow' than a butterfly style. The arrangement of body and actuator gives less linkages normally (direct coupled), but this can require more space in some directions, which may require location to be considered early in the project planning. Again if fibre is present in any quantity this could cause blockages in this type of valve, but being of a hard seat design it is better suited to high pressure drops and slurries, etc.

Paper stock and slurries   Where high levels of fibre, etc. are present, e.g. paper stocks, the profiled ball or rotary plug types are best suited. For some applications, e.g. stocks above 2% consistency, only the profiled/segmented ball and rotary plug are suited if blockages, etc. are to be minimised and avoided. Some special types are available, e.g. **basis weight** control valves, for paper stocks and slurries. These are highly accurate and very expensive, but are essential for the special application for which they are designed. Generally they are very sophisticated profiled or segmented valves.

Fig. 3.8 DeZurik precision electric basis weight paper stock valve. The positioning of the rotary segment in this control valve is highly accurate, and is used where extreme accuracy of flow control is required, such as in the 'basis weight' on paper machines of all types.

Gas and steam fluids   Again the butterfly types can be used on flow control duties, but particular attention has to be paid to temperature and pressure drop where 'soft seats' are used. For high temperature or high pressure drops the 'soft seat' will not generally give good life or will be totally unsuitable. For steam service of all types globe or plug and seat are normally the best suited. For special applications profiled ball types are available, but they carry the penalty of higher price compared with globe for similar duty applications, since the seats have to be of special materials to withstand the temperatures. In the larger sizes, however, the profiled ball may give a better characteristic for the application, and be less of a differential in price to the similar Cv globe since the globe may need to be of bigger size for a similar Cv.

Powders   For these media temperature and pressure are not usually a problem, but the medium limits the selection of valve types to pinch, spade and some butterfly types. In some applications ball types can be used but generally have a high cost penalty. Other special types have been developed for powders, but these are generally included in a package of specialised handling equipment. For many powder applications variable speed blowers are now more attractive than control valves.

Sludges and slurries   In most applications these will fall primarily into the category of paper stock, but some may be more similar to powders, e.g. the use of variable speed pumps being more attractive now.

Foodstuffs   All the above will apply according to the medium type with the addition that the internals, at least, must be sterile and capable of regular sterilisation. Generally, as a minimum requirement, body and internals will be stainless steel. Butterfly and ball/segmented ball type valves are most suited to this application. The design used should be such as to minimise areas where the medium could build up or lodge.

We can summarise the valve body type suitability as follows.

1. *Butterfly*   Good general-purpose valve, low cost, limited application where valve modulates below 20% or above 80% opening. Most suited to 'clean', low viscosity fluids. Generally will not give repeatable tight shut-off on hard seats, e.g. if used on steam, and is not really best suited for pressure control applications.
2. *Globe*   Most suited to gases and fluids where high temperature and pressure drops may be encountered. Can give tight shut-off if required, by careful lapping of plug and seat, and is hence very suited to pressure control applications. Especially suited to steam duties.
3. *Ball*   Most suited to high fibre content and viscous fluids, etc., especially applications where repeatable tight shut-off is needed. For accurate control profiled/segment ball, vee-ball, etc. should be considered. Ball type valves will normally give the best, most repeatable shut-off of all modulating valves.
4. *Spade or knife gate*   Not very suited to control applications other than on-off service. Even vee-notch types are generally considered unsuitable for continuous control purposes today. However, fitted with air

cylinders the spade valve makes a very cost-effective on/off valve. Two or more fitted with cylinders are well suited to fluid diversion control.

5. *Pinch* Not found on many control applications other than powders, etc. and where used do not often give very good modulating control.

The above details are by no means definitive and there may be occasions where the types of valve body applications will differ. In such cases, providing all the required properties of control are satisfied, there should be no need to change the unit installed or proposed.

One valve not detailed above is the **three port**, **splitting** or **diversion valve**. This is a special-purpose valve and is used for special applications only. It is generally very expensive and for many applications a pair of cylinder operated spade valves will serve the same purpose. Only if the process/fluid type means the spade valve combination is not suitable will the expense be justified. The process design engineer should specify the requirements in such cases.

Control valve actuator

We have covered the valve body in some detail so now we need to concentrate on the device fitted to the valve body to give the modulating action required for control purposes. This unit is generally termed the actuator. All the valve body types require to be fitted with an actuator to modulate the valve to an accurate position according to the applied control signal. The actuator can be of many differing types, as mentioned at the beginning of this chapter, but still the most numerous today are the pneumatically operated types. Many actuator types designed for continuous control can be applied to cyclic control applications by modification or addition of converters, etc., e.g. solenoids. With *all* control valves the quality of the fitting of the actuator to the valve body is extremely important. *If any slop or backlash is present the quality of control will be impaired.* In situations where large amounts of backlash are present, poor, unstable, or loss of control can result.

Pneumatic actuators

Pneumatic actuators generally come in two main types, namely **diaphragm** or **piston**, although some do have 'air motors', etc. The diaphragm designs are normally the lower cost, and have better durability, but the piston models can provide higher powers and have very high power to size ratio, provided adequate air supply pressures are available. The sizing of the actuator for the application is dependent on the thrust required to operate the valve under *all* the service conditions to be encountered. These conditions will have been identified from the data entered on the control valve specification sheet. One piece of data that is often overlooked is the 'D.P. (differential pressure) closed'. It can sometimes be critical for the valve to remain closed when the parameter under control is at minimum value, e.g. pressure control on a vessel when the supply is at full pressure but pressure in the vessel is zero. In this condition the D.P. closed is the maximum process supply pressure! This condition can change the size of the actuator considerably in some cases, and may have implications with respect to the *safety requirements. A control valve should not be relied upon to give isolation of plant,* but if the valve will not hold shut under all conditions it could cause process or possible safety problems. Providing it will remain shut under all conditions it will satisfy the process requirements, complement the isolation valve, and hence add to plant safety. If, for process and/or safety reasons, the valve must close, hold or open when modulating power is not available, this must be specified on the control valve specification sheet (*air fail action*).

Diaphragm actuators   These tend to be fairly large where high thrust is required which can give rise to location problems on limited space applications. Especially where high D.P. closed conditions exist relative to the working range D.P. the diaphragm can be one or two sizes up relative to normal control conditions to cover this situation, especially with globe type bodies. This will mean that the volume of air required to move the **valve stem**, or **spindle**, could be quite considerable, resulting in reduced response to control signal changes. If the D.P. closed gives rise to a larger size actuator the volume of air required under normal operating conditions may result in the speed of modulation, and hence control response, being reduced significantly compared to the actuator size to cover normal control conditions. This limitation may be further aggravated by the controller or signal converter not being able to supply adequate volume of air to satisfy the control valve actuator demand. To overcome the limitations a **positioner** should be employed (the positioner is described below). This device not only has the capability to increase the air volume at the valve actuator, necessary to operate the control valve at the greatest speed possible (providing the air supply is adequate to the positioner), but can also be used to 'characterise' the valve position to the control signal. Should the control valve need to be used on cyclic or on-off control the actuator will require fitment of a solenoid valve to satisfy the duty, but the body and actuator sizing constraints detailed above will still apply. The fitting of a positioner may also be advisable to meet response speed requirements in the case of cyclic control applications.

Piston type pneumatic actuator   This has the advantage in many applications of its small size relative to the diaphragm type for the power available. The disadvantages, however, are the increased wear surfaces, springs necessary for some air fail states and types of action, high pressure air supply and higher cost than diaphragm types. In the 'double acting' piston types the response speeds can be very high, but in all forms the piston

Fig. 3.9 Fisher control valve in a typical oil refinery/chemical plant environment. The actuator in this view is a pneumatic diaphragm type (courtesy of Fisher-Rosemount).

actuator must be fitted with a positioner for modulating duty. Some piston types have an additional bonus in that they can be provided, like diaphragm actuators, in a form to give 'rotary action', which is extremely useful when applied to rotary valves such as butterfly, vee-ball, etc. Due to the high forces which can be present on rotary action the quality of fitting of the actuator to the valve stem is extremely important, since any backlash will soon result in rapid wear, and early failure. These actuators are very useful for applications such as combination fuel/air valves in burner control, etc. If this type of actuator is proposed by a supplier, check the air supply requirements since some do require lubricated air. If this is the case also check the types of lubricant permissible, as many rubber components will be damaged by some oils and greases. Also check the suitability of the positioner to accept lubricated air. The 'double acting' piston types, fitted with a positioner, are also very useful for actuation of modulating dampers on heating and ventilating ductwork, etc.

Control valve positioner

As mentioned above, to ensure the quickest response of the control valve the fitting of a positioner should be considered. A positioner is effectively a pneumatic amplifier, utilising a local air supply for the volume/pressure requirements and balancing the signal air pressure to the air supplied to the actuator so as to maintain actual valve stem position by applying more or less air pressure as needed. Not only will a positioner enhance the response of the control valve to signal changes, it will also ensure repeatable positioning of the actuator, hence the name, and will enable the control valve to be characterised for the application. Characterisation allows conditioning of the control signal to give a characterised throttling action by the control valve, e.g. linear control signal converted to 'square law' throttling, etc. As you will no doubt appreciate, the use of a positioner can be applied in another way to make a control valve 'linear' in its relation to the control signal, which is generally the most common function. This will allow a control valve to be easily characterised to the controller to give improved control over the process variable under control where non-linear performance may exist from one part of the valve operating range to another. The three most common control valve flow characteristics are **linear, equal percentage** and **quick opening**. The selection of which type of characteristic is best suited to the application can be summarised as follows.

1. *Linear* Gas process of large volumes, long pipes, large vessels, etc. where D.P. decreases with load increase (change in D.P. > 20%).
2. *Equal percentage* Liquid process and gas process not covered by the above conditions. Possibly the most common characteristic used.
3. *Quick opening* Where D.P. increases with load in situations where the change is in excess of 200% from minimum to maximum, or where emergency dumping of process fluid is required.

Some applications may require different characterisation to the generalisations above. In such cases this can be done by making special cams for the positioner or, if microprocessor-based controllers are used, the output of these can sometimes be characterised.

Another common use of the positioner is to allow 'split-ranging' of control valves to be achieved (e.g. 50% of signal opens valve 1 fully, and the next 50% opens valve 2, etc.). In theory split-ranging can be done on several valves with one common control signal, but in practice the number should be limited to a maximum of 3 (in special cases 4), due to the possible drift in calibration which can occur over a fairly short time in service. Today positioners are available with inbuilt electro-pneumatic converters (I to P or E to P), but pay particular attention to the vibration possible at the control valve. Ensure that the complete unit will withstand all the vibration likely to be encountered. The design of a complete unit may save installation costs, but incur running problems which can only be overcome by installing a separate signal converter, away from the vibration environment. If vibration is expected detail it on the control valve specification sheet. Remember the earlier quotation: '*You get what you pay for; if it is not specified, it will not be included*'. It still applies!

With all pneumatic actuators ensure that the air supply, both control and power, is of the highest quality. *Dirty air means poor control, low reliability and increased maintenance.*

Signal converters

Where a controller output is electronic a **signal converter** (I to P) will be required to give the pneumatic signal necessary for the pneumatic actuator/positioner. Some positioners encompass the necessary signal converter within the unit, as mentioned above, but the vibration likely to be encountered should be carefully considered when selecting such devices. It is always worth locating the signal converter as close as possible to the positioner/actuator to give the maximum signal response. For applications in hazardous areas the use of correct housings, barriers, etc. must also be observed, or locate the signal converter in the 'safe' area and tube the safe pneumatic signal through the hazardous area.

3.2 Variable speed drives

Another control device is the **variable speed drive**. As the title suggests the speed of the device is changed in direct relationship to an applied control signal. These control devices can be either electrical or mechanical, both giving similar results. The mechanical types are generally of a variable ratio gearbox, but tend to be limited to smaller power sizes, and of limited turn down. The mechanical units are of particular use where the prime mover is not electric, where low speed ranges are required or where the controller output is not electronic, e.g. pneumatic or hydraulic. It is quite common to find a pneumatic actuator fitted to a variable ratio gearbox to achieve a variable speed drive. By far the largest and most common variable speed drives today are the electrical/electronic types. These are particularly useful on centrifugal pumps, etc., and on large power applications can effect considerable power savings over control valves and fixed speed pumps, etc., especially where a large amount of load/speed changes are necessary to maintain control. Most of the smaller units available today are 'inverter' type, whereas the large power units are quite often dc (direct current). With technology development variable speed ac drives are becoming more common for large power sectional drives. The accuracy of speed control of these units is now comparable to their dc counterparts as used on paper machines, rolling mills, etc., and they generally have a price advantage. There are no hard and fast rules governing the selection of the type of variable speed drive for any application, except to ensure that all conditions likely to be encountered by the drive will be well within its capabilities. Many drives will not cope with continued running at very low speeds, unless provided with forced cooling/ventilation. If long periods of running at one speed are to be encountered it may be more economic to use a fixed speed unit and a control valve, since the cost of the drive may be much higher than the valve, and hence may never give a full payback when all the maintenance costs are included in the 'total cost' appraisal. Variable speed drives are not well suited to many pressure control applications, and will not have the control valve advantage of being able to 'hold a pressure' in a vessel unless a non-return valve is fitted or the pump is of a 'positive displacement' type. Variable speed drives are at their best on flow type applications, whether this be on pumps, machines, production lines, etc. On applications where the signal-to-speed relationship is non-linear, or of the wrong characteristic to suit the application, any characterisation must be done in the drive electronics or at the controller output. This should be borne in mind when selecting equipment since this feature does not always come as standard on all units, either controller or drive. Remember; *if it is not specified, it will not be included.* Additionally many variable speed drives available have an in-built PID (proportional plus integral and derivative) control function. This can be used to considerable advantage if only one discrete loop is needed, or there is no need to centralise the control functions. This facility could also be used for the flow loop control as part of a blend control system, the flow control set point being generated by the blend controller.

Some variable speed drives require a considerable time for starting, so ensure this can be tolerated in the control strategy of the process under control. In many cases extra I/O may be required to satisfy the interlocks needed by the drive, as well as the extra data for display, etc. On fixed speed electric motors a minimum of one contact input and one contact output is required, but on many variable speed drives a typical minimum of two contact inputs and two contact outputs is required. The choice of control device can considerably influence the I/O needed for control purposes, and therefore may also limit the scope of choice of control equipment as a result. With so many different manufacturers and types of drives available today it is necessary to draw up a good specification of what is required of the drive, including all the running conditions to be encountered, and present this to all the prospective suppliers. The greatest deciding factor may be as simple as a current site standard, the supplier who gives the best support for your site, or the one that can be accommodated by the available I/O on the control system.

On any variable speed drive ensure that the control signal failure mode is specified to suit the application, since some drives will run at maximum speed, others at the last speed signalled and others will stop on disconnection of control signal. Where the drive requires a pulse signal to change the speed ensure that the controller can deliver the correct pulse rate, voltage, etc. Also check if the signal cable requires screening or special cables as this will influence the installation of the drive, possibly in time and costs. If there is no specification for the signal cables it is best to use a screened type to ensure that no interference will be presented to the unit. Before final selection of any electrical variable speed drive it is worthwhile confirming agreement with the site electrical department (and the mechanical department for a mechanical unit) who may be ultimately responsible for the maintenance and support of the unit.

Another form of variable speed drive, already mentioned, is the **vari-**

able ratio gearbox. This is a mechanical device fitted to a constant speed motor to achieve a variable speed at the output shaft. Generally these units are fitted with an actuator of some type, to achieve the change in gear ratio. It is this actuator to which the control signal is applied. There are also some other types of variable speed devices such as **magnetic flux couplings**, and 'fluid couplings', to name two, which can give a variable speed output from a fixed speed drive input. Generally these are not very common and, due to their complex nature, need a high level of maintenance.

3.3 Directly coupled actuators

Most directly coupled actuators are supplied as part of a piece of equipment. It is essential that the control signal standard, used on site, be specified to the supplier prior to any order being placed. In some cases a change of supplier standard to match the site standard may be difficult or even impossible. In such cases a signal converter should be used to convert the controller signal to suit the directly coupled actuator. Where the signals are electronic, or digital, it is always advisable to employ a converter giving full signal isolation between input and output, to overcome any compatibility problems, e.g. avoiding 'earth loops', etc. There is little more to say about direct coupled actuators since any detail is generally down to the supplier. One example of directly coupled actuators is the Thermal Rod™ used by Measurex Corporation for profile control of a paper machine slice where expansion and contraction of an actuator rod is the power of the device. Another example is the Modutrol™ actuator used for modulating burners on many industrial boilers.

3.4 Other types of control device

One of the other control devices not mentioned so far is the **variable stroke metering pump**. These are often encountered on applications involving chemical dosing/injection into a process, e.g. pH control, boiler feedwater treatment, etc. These are effectively a variation of the variable speed gearbox. Other types of control devices include 'stepper motor' type units, linear actuators, hydraulic actuators, etc. In all cases ensure that a full specification of what is required of the unit, the operating environment and the signal standards to be used are included in any enquiry for supply, and include a request for full details of signals, connections, power requirements, etc. to be supplied with any proposal, to ensure that the unit will match the application. In this way a trouble-free installation can be achieved. As pointed out in the section dealing with Mass Flow Measurement in Chapter 2, the effect of pump pulses can cause problems with controller tuning, if the measurement device is sensitive enough to detect these. In such cases an input filter (PV damping) will be necessary on the controller input to average the flow measurement. In most cases, however, the flow rate of the pump derived from the pump data will be of adequate accuracy, and flow measurement/control of the pump discharge will not be necessary. The pump used as a final control device should also render such flow control unnecessary in most applications.

There are several other types of control device, but most of the commonly encountered types have been covered here.

With all control devices the applicable regulations and guidelines for the application must be observed, e.g. pressure vessel regulations, etc. and additionally the applicable regulations and guidelines for hazardous areas must be strictly adhered to, wherever installation is to be made in a classified zone.

4 Cyclic (on/off) control

Let us retrace our steps a little and once again consider the type of control we require. From our investigations carried out in Chapter 1 we should have identified the type of control required for the application. In this chapter we will consider the outcome of the investigations as having identified cyclic (on/off) control. We have looked at the measurement devices in some detail, so now let's see if the different units covered so far can be used for cyclic control.

4.1 Measurement devices used for cyclic control

In many cyclic operations the measurements are very basic, e.g. temperature switch, level switch, etc. or there may be no measurement at all. Take the example of the cistern where the process cycle is initiated by pushing or pulling a handle, not by a measurement. If we consider a sump drained by a pump, the process is cyclic, the cycle started when the level goes high and stopped when the level goes low. Here we can use a level switch to start and/or stop the pump, the level switch being a measurement device. In every case the environment in which the device is operating must be considered, not only external, but also the process medium, as mentioned in Chapter 2. The process medium is the prime consideration since plant and personnel safety will influence the choice of equipment, as well as whether the measurement device will be suitable for the general working environment. To ensure that all aspects are covered it is worthwhile using the transmitter specification sheet for all measurement devices for any type of control, always answering all the items on the sheet for each device and entering N/A where the section is not applicable.

For basic cyclic control the measurement device will often need to be only a switch type device, e.g. level switch, temperature switch, etc. and this information can be included in the standard transmitter specification sheet.

Fig. 4.1 Trip amplifier by Camille Bauer Controls Ltd.

In some cases the measurement may be carried out by an analogue device (transmitter) which gives an output signal directly proportional to the value of the measured parameter, as described earlier. The analogue output signal is then fed to a device giving a switch function at the desired value. This device is known as a **trip amplifier**, **signal converter**, etc. The transmitter specification sheet will apply to the analogue measurement device and an additional sheet can be used for the signal converter, the calibration being the input from the transmitter and the output signal standard being the switch outputs required. The trip amplifier can have many switch points and can therefore be used for multiple functions from one transmitter, but if required these will have to be specified. Alternatively a separate **signal converter specification sheet** can be drawn up and used. For such a specification sheet only the applicable sections need be transferred from the transmitter specification sheet.

If you were setting up a small sump level control system to start and stop pumps, and give alarms for high level and critical high level, one **level transmitter** connected into a trip amplifier, having several output switch functions, could satisfy the control requirements easily (Fig. 4.2), whereas you would require at least three separate level switches to achieve the same result. This combination of transmitter and trip amplifier has an added bonus in that should other devices be required, e.g. a continuous record of the sump level, a recorder can be connected in the circuit between the transmitter and trip amplifier to give this function. The recorder, etc. can always be added easily at a later date if the requirement is not initially identified.

As you can see from the above consideration of the functions of the controls required, and other possible requirements, in relation to the measurement, the resulting specification can quickly identify the nature of the front end of the loop/system design. Any system which has many components can be difficult to understand and maintain, and can also be very difficult to fault find should an item fail, so if there are more than two actions to be taken against a measurement, consideration should be given to a transmitter/trip amplifier combination instead of a number of switch type devices. In Chapter 6 we will look at all the above functions being further rationalised in controllers, etc.

We looked at the different types of measurement devices/transmitters available in Chapter 2, all of which can be applied to cyclic control using signal converters/trip amplifiers as detailed above. Now let's go to the next element in the control loop, the **controller**.

4.2   What type of controller?

We have covered some aspects of measurement and touched on control in the form of a trip amplifier. Now let's expand on the subject of the con-

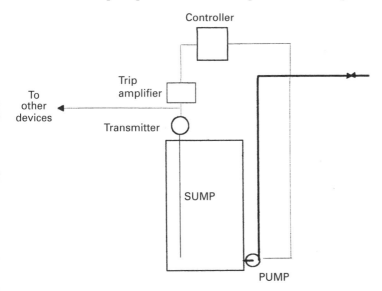

Fig. 4.2 Typical loop diagram for sump level control (using a transmitter and trip-amplifier).

troller. If we used a level switch in the sump control system as shown in Fig. 4.2, the switch operating the pump must be capable of having a wide **dead band**, otherwise the pump would be switching on and off continuously as the level bounces, giving rise to damage to the pump, motor and switchgear, etc. There are some types of 'float' level switches with this dead band facility inbuilt, as shown in Fig. 4.2, manufactured specifically for this type of application. In Fig. 4.2 the same control can be achieved with the added bonus of other functions, such as alarms. These would need a small amount of dead band to stop 'fleeting', otherwise the alarm would follow surface ripple, etc. To overcome the pump switching problem two level switch points can be used to switch the pump on at high level and off at low level. Sump level switches are available giving this exact function, Fig. 4.2; but with other, additional, requirements, such as alarms, perhaps consideration should be expanded to other control devices such as **pump controllers**, or programmable logic controllers (PLC). Where a number of similar control tasks are required in an area of plant, careful choice of control equipment can achieve considerable cost savings as well as improved reliability and lower maintenance costs. For example, networked or multiple pump controllers or a PLC covering an area of plant could be chosen. If the control system under consideration is truly stand alone, a simple control device is all that is required, but if the control forms part of, or is likely to form part of, a much larger scheme, then the overall scope should be taken into account. A simple clue to the potential control scheme size can be obtained from a **flow sheet** of the process, and/or whether the pumps, motors, etc. will be connected to a multi-motor panel (MMP). With current practice where possible MMPs are used to segregate electrical apparatus from the process environment, improve reliability and plant safety and aid maintenance. This electrical apparatus location can assist the process control engineer since early planning of equipment location can save considerable cabling costs by locating the controller with, or close to, the multi-motor cubicles. This type of planning will obviously influence the control equipment type, etc. so reducing the scope and making choosing easier. With modern MMPs, however, many have the facility to transmit and receive data via communication networks using twisted pair or co-axial signal cables, which can hence reduce cabling requirements considerably. If such 'intelligent' MMPs are used specify all the details to the potential control system supplier. It is worthwhile ensuring that the intelligent MMP installed will communicate directly with the proposed control system without the use of a special interface. This will reduce the number of items to give potential problems and reduce the potential snags when software is up-dated, etc.

Where simple pump controllers are employed the equipment should be located where it best suits the way the plant is to be operated. Where the scope of control is mainly on/off, cyclic or timing type control, the first choice should be for a PLC or a network of PLCs. PLCs can be operated by purpose-designed control panels, or via a video terminal and keyboard. However, some of the low cost units are expensive to interface to video terminals, and also difficult and expensive to network. If considering a particular PLC investigate its suitability for potential control expansions. Initially, specifying a more powerful unit capable of easy, low cost expansion, can quite often save much cost and embarrassment later! It is also worthwhile considering networking a series of standard size units. This will standardise spares holding, make fault finding easier, can be sized to suit process plant modules and can often save cable costs, etc. There are many PLC manufacturers about today and the choice can sometimes be daunting. The choice can be reduced by considering a few simple factors. Your answers to the above considerations will have already reduced the choice somewhat, but additionally you should consider what has been used on site before and whether there is a site standard. If equipment has been used on the site before, has it been reliable? Has it been well supported by the supplier? Are the existing units on site still of current availability, or have they been superseded? Will the site units fit this job specification? If there is a site standard, and any of the above questions do not have a positive answer, should a change be considered? Before proposing a change to an existing site unit, the site usage of the equipment must be considered, since a high number of installed units should have the back-up of spares, site experience, etc. You should therefore be able to obtain advice from site personnel familiar with the equipment.

4.3   Systems houses

If the scope of the project is quite large, and specification generation and control equipment application is not your regular function, plus there is nobody to assist you, do not despair, help is available. Many **systems houses** are available. They specialise in PLC application and are able to propose, estimate cost, design and install PLC equipment. The choice, however, is down to you. So how do you identify the most suitable systems house for your project?

Systems houses can generally be split into two categories.

1.   Those specialising in a particular industry.
2.   Those specialising/aligned to a particular range/manufacturer of PLC.

Both the above types of systems house may be approved by the PLC manufacturer they use, but each has its advantages and disadvantages. For the first application in your plant the industry specialist may be the better choice, providing the track record is consistently acceptable to existing clients. If possible contact some of the previous clients to get an idea of how the systems house performed. On the other hand, if there is already a considerable site usage/experience of previous PLC installations, a specialist in a range of PLCs can probably offer in-depth experience of that range to get the best functionality from the equipment.

   To find out details of local or industry specialist systems houses the easiest way is to contact several PLC suppliers for details of their recommended/approved systems house. Most of the 'trade' magazines carry a large number of advertisements for PLC manufacturers, as well as being able to provide a comprehensive data-base built up from months of magazine issues. If this course of investigation fails your local library can usually provide some information.

Before approaching any systems house ensure that you have developed a very detailed specification of the process control requirements for the project in hand. We mentioned flow sheets earlier. These are the start points of your process specification, but much more detailed information is required to give any systems house a chance to quickly understand what you require. A little time spent on generating a well detailed process control specification will pay high dividends.

4.4 Process control specification

We have covered the measurement device earlier, so for each control loop the measured input can be defined (see applicable specification sheets). Also the various functions associated with each measurement can now be defined. Let us now draw simple loop diagrams for all the loops, showing the controller as a box. If we now take all the loop diagrams associated with each area of process plant and put them together, we will quickly build up a full picture of the control requirements in each area. For now we should consider the control requirements not the plant geography; this may, however, influence the arrangement later. We should now have several control boxes for each plant module. Alongside each of the control boxes we should be able to define what each does, process wise, in plain English, e.g. level less than 20% stop pump, level greater than 70% start pump (lvl < 20% pmp stp ; lvl > 70% pmp strt), etc. From this information the systems house can gain a fairly good understanding of the control requirements and give a very reasonable estimate of costs, design time, etc. But this costing will only be for the 'controls part' of the project. The systems house will also be able to advise on PLC type, how many units, how they should be networked, etc. However, there is still more information that should be included in the process control specification if a total price, including equipment, programming, etc. is required.

1. How the plant is to be operated, e.g. totally automatic, central control room, purpose designed panels, etc.
2. Layout plan of plant showing location of each measurement device.
3. On layout plan show the location of MMPs and other switchgear, etc.
4. Site working voltages.
5. Where PLC equipment could be located.
6. Identification of any hazardous areas where equipment is to be applied.
7. Copy of the transmitter specification sheets.
8. Copy of the control valve specification sheets.

9. Copy of any plant operation details to be interfaced/controlled by the system.

From the information detailed above the systems house can estimate the ideal sizing of PLC units to suit the plant and the operation. The plant layout, 1 above, will allow the systems house to take the plant geography into account which will influence cabling, equipment location and maybe PLC unit sizes. Also much of the basic operation information is included giving a good indication of the complexity of the PLC programme required. Remember; *you get what you pay for*, and *if it is not specified, it will not be included*. For example, if you want full programme documentation this must be specified with the enquiry. The more requirements you can detail and include with the enquiry, the more accurate will be the quotation, as well as all the required equipment proposal and cost estimate.

4.5   Control devices for cyclic control

The control device is the item that 'closes' the loop and carries out the necessary action on the process desired by the controller to keep the process under control. In the case of the sump control example earlier, the pump is the control device. In practice, however, it will normally be the electrical switchgear associated with the pump, combined with the pump, unless the pump is small, single phase and of low voltage electrical operation. In other cases the control device may be an on/off valve, solenoid, relay, etc. The control devices covered in Chapter 3 can generally all be applied for cyclic control with modification such as solenoid valves, relays, etc. The process design and equipment will normally define the control device for the process, and this must be specified in the enquiry documentation to the systems house, including the device specification and who will be responsible for its supply. This information is included above in items 7, 8, and 9, and should give the systems house all the information they will need to propose the correct equipment.

So far we have established the items required for on/off or cyclic control type applications for our process, along with where we can get assistance for pricing, design, etc. The following chapters will develop the complete control system, and enable the reader to specify one, or to assess what is being proposed by a supplier and which type of control system will best suit the application. We have identified that cyclic control can be catered for by use of PLCs, supported or supplied by a systems house, but the control requirements can also be satisfied by other systems such as a distributed control system (DCS). In the next chapters we will endeavour to establish the major differences between the systems, and which of the control system types can be economically applied to the process being automated, along with the necessary considerations needed when applying the control systems to the process.

5 Programmable logic control (PLC) systems

In the previous chapter we looked at cyclic control and identified where help could be sought for application of programmable logic controllers, so now let us investigate PLC in more detail. The PLC was initially developed for the automotive industry, where the automation of the continuous production lines required a controller capable of carrying out the basic repetitive/sequence type control operations on the production lines. Relay control had been used for many years for these duties, but had the disadvantage of large panels, heat build up, difficult fault finding and inflexibility for the production line. The need for relays to be 'hard wired' resulted in very large, complex panels, and the pure nature of their assembly severely limited control requirement changes, whatever the size. The development and availability of 'solid state' technology was the answer to the problem, and the PLC was born. From the mid 1970s PLCs developed very rapidly for the automotive industry and quickly found use in other applications where similar operations were required, e.g. packaging, machinery, etc. Their application to automotive assembly, packaging machinery, etc. guaranteed their rapid acceptance for these applications, and hence ever increasing sales. They remained sequence type controllers for many years, the handling of analogue signals being rather unwieldy and awkward, but basically unnecessary for their designed use. As the original market place became saturated other markets were sought and other applications found which required more functionality, and the need for processing analogue signals. The analogue processing remained limited until the introduction of microprocessors in the mid/late 1980s.

The availability of smaller and more powerful processors, and then microprocessors, allowed size reduction and increased capability of all PLCs. The smaller units still remain basically sequence controllers, but the larger units were soon capable of handling a reasonable level of analogue control, along with many other facilities. There are now two distinct types of PLC avail-

Fig. 5.1 Small PLC suited to small sequence control systems.

able, the small, stand-alone limited function, low cost PLC, and the powerful, networked, multi-function PLCs. We have mentioned the example of a pump controller, which is effectively a stand-alone small PLC, having limited functionality but very cost-effective for the designed application. Some of these small units have limited communications available for display/monitoring, etc. and are not cost-effective for integrated control systems and the like. Where local stand-alone control is needed, with little or no information required at a control room, etc. the small, low cost PLC is a very good solution, but where any degree of centralised/remote control is envisaged, the networked types are definitely the prime choice. With advancement in technology the cost per point of I/O has also decreased, making the PLC, whether small or large, an economic alternative to relays, when only four or five relays would suffice for the control requirements.

5.1 Operator interface

PLCs on their own do not have an operator interface or display as standard operating of any control system employing PLC control will require some means of inputs and outputs to allow the operator to action controls and get feedback of control status. Many have I/O status indication, but this in no way should be regarded as any form of operator control inter-

Fig. 5.2 Siemens SIMATIC™ PLC modules. Such modules allow various input and output types to be accommodated, including both switched and analogue.

face, and does not give a means of inputting control functions into the PLC. The necessary inputs and outputs can be as basic as push buttons and lamps, but increasingly video terminals and keyboards are being employed. Some are limited in what is displayed, some being single, or multiple line displays, with dedicated keys, and the ultimate is a PC-based full colour graphic terminal and keyboard. Where such interface terminals are used the display is generally generated using a SCADA (Supervisory Control Adatacompnd Data Acquisition) package of software to generate the graphics, which can give the interactive interface between operator and PLC. We will be covering SCADA packages in Chapter 8.

5.2 Programming

PLCs are programmed by using a specially designed programming device, or in some cases a PC, communicating via a modem, or similar communi-

Fig. 5.3 Siemens SIMATIC™ PLC data/control terminals.

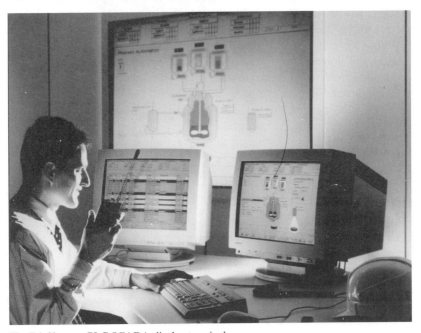

Fig. 5.4 Siemens PLC SCADA display terminals.

cating device, running dedicated special software. Most are programmed in 'ladder logic' or 'flow logic', in a pre-structured format, and in general the more powerful units require a training course to enable the user/owner to programme or maintain the unit. For fault diagnosis, programme change, etc. a programming device of some description is essential, and generally changes can only be made when the PLC is off-line, although on-line operation can be observed on most units. On some units minor changes can be made on line, such as forcing the state of inputs or outputs, changing tuning parameters, etc., but nothing more in the running/on-line state. It is worth pointing out that many programming devices require a power supply independent of the PLC, so it is worth ensuring that such a supply is available local to the installed unit. PLCs do have the advantage, over many other control types, that their operating speeds can be extremely fast, but only a few units available today are 'multi-tasking'. This has the disadvantage that if there is a programme or field device fault it can 'freeze' the sequence. For this, and for safety reasons, the programme should be written to include a check of all switch functions, etc. to ensure that each step is carried out properly, no false conditions exist, and that failure of any stage of sequence or field device will raise an alarm and not allow a sequence to start or continue. Like all software-based systems programming must be structured to ensure reliability, easy fault diagnosis and the best functionality from the PLC. Programmes which require multiple scans to complete a sequence will give rise to slow operation, possible 'lock-ups' and waste of capability of the PLC. It is always advisable to draw out a flow chart of the sequence required to ensure that all the steps will follow logically and to identify the capacity/size of the PLC needed to satisfy the control requirements. When programme modifications are carried out a copy of the programme prior to alteration should be made and all redundant parts of the programme should be removed. Good house keeping of PLC programmes will assist trouble shooting and avoid faults due to unwanted/redundant programme parts interfering with the running sequences. To minimise confusion and save time when fault finding, it is worth dedicating each field device, e.g. switches, solenoids, etc., to its own input or output on the PLC. This may increase I/O requirements and cabling, etc. but will ensure the fastest fault finding possible, should a field device give problems. It will also allow any field device to be used in any part of the programme, whenever it may be necessary, without modification to wiring, hence saving time, and giving a much more flexible control format. Dedicated I/O will also allow field device over-ride with the programming unit during fault finding, without the need to touch the wiring, but *this facility should only be employed for fault finding under carefully controlled conditions, and never used to over-ride devices on a running plant.* Good programming practice, such as identifying every part of the programme with text and describing the function or sequence, will aid fault finding at a later date even by personnel not familiar with the original programme. It is also worth considering including a function check of all switches used in the programme, ensuring the switch state does change, and is not just observed by the programme to be in the correct state for a sequence action, at the start of that sequence.

5.3 Stand-alone PLCs

The smaller stand-alone models normally come fitted with all the I/O terminals needed to utilise the full capacity of the unit. Many do not have communication facilities to allow operator control, etc. via a video type terminal, so I/O required for operator functions, status information, etc. must be included in the I/O count when designing/sizing the unit. When applying such units to a process attention should be made to the I/O rating for both voltage and power. On many of these units the total power capacity needs to be observed carefully since this may not be a straight multiplication of the individual I/O ratings. The environment average and maximum temperature will also affect the overall rating of the unit; since most PLCs do not have fans fitted, it is essential to observe the manufacturers' recommendations for load versus maximum operating temperatures, and for the type of enclosures to be used to house the PLC. In some instances there may be no additional cooling requirements. In others just a circulating fan may be needed or, in high loadings/high working temperatures, some form of cooling may be required. Due to the termination styles used, all PLCs should be housed in electrical enclosures, or purpose-built cabinets, capable of being locked. This puts an on-cost on any PLC installation and needs to be considered when designing the system. Even the small stand-alone units need to be housed in a manner which meets the electricity and safety regulations, as well as providing protection from environment problems. The size and type of cabinets, etc. will affect the working temperature of the units housed within them, and the possible resulting temperature may necessitate limiting the power rating of the PLC as outlined above. All PLC manufacturers publish full data, including maximum loadings for their various PLC models relative to working conditions, and the manufacturer, or main supplier/distributor, will give advice on the best way to ensure reliable working conditions for the equipment.

If using a system house to propose a system for your application they should be able to carry out all the necessary calculations to ensure that the correct size PLC, cabinets, fans, etc. are included in their proposal. The proposed/preferred manner of housing of the PLC should be detailed to the system house with all enquiry documentation so as allowances can be made in their proposal. With any proposal it is worth asking, 'what will be the temperature rise above ambient within the cabinet?' and checking the resulting maximum working temperature against the PLC manufacturer's specification. Consideration should also be given to the seasonal temperatures likely to be inflicted on the installation, plus what air movement is in the area. This may seem very basic but often a high working temperature causes many problems with equipment, sometimes resulting in failure of the PLC. Also ensure that any cabinets/panels are suitable for the working environment; a wash down with a hose is not conducive to good PLC operation! Cabinets (enclosures) and panels will usually comply with an environment rating such as IP, NEMA, etc. and the rating code will specify the conditions the cabinet will endure. Most enclosure manufacturers will give details of the coding/rating on request, and many publish details in their catalogues.

We can now generate a small check list to ensure we correctly specify or confirm a proposal of what is required. This can easily be done by answering the following questions.

1. How many inputs, voltage and field device type?
2. How many outputs, voltage, load of each, field device type?
3. Specification of PLC I/O, e.g. non-isolated, and total load capacity.
4. Capacity of PLC required?
5. Size and type of enclosure required?
6. Maximum working temperature expected?
7. Specification of maximum working temperature of proposed PLC.
8. What operator facilities are required?
9. What communications are required?
10. What type of programmer is required and what power supply does it need?
11. Hazardous areas?

5.4   Large and network PLCs

Like the small stand-alone units these require to be housed in cabinets/enclosures or, in cases of a large number of units, perhaps dedicated rooms. Again the working temperature needs close attention at the design stage to avoid problems of overheating when the PLC is working. The architecture of these models can, however, differ from the smaller units, with many designs now available in a modular style of architecture where the I/O and processor are built up in a series of modules allowing a very flexible, large system to be developed. Figure 5.5 shows a basic design of networked PLCs where they and the I/O modules are housed in cabinets installed in locations to suit the plant. The modules available allow the PLC to be 'tailored' to the application and, in many cases, allow easy expansion for future projects. The expansion of these units is, however, limited according to the I/O make-up, programme size and/or complexity relative to the memory size of each unit. The unit/module capacities available will satisfy most complex, integrated sequence requirements and the networking of units allows the largest of projects to be accommodated. Even with many PLCs employing modular construction the make-up of I/O power loading must still be considered to avoid over-loading a module. The considerable advantages of modular architecture over the fixed strategy of the small stand-alone PLC makes these units an obvious choice when a 'flexible' PLC system is envisaged. The modules are each generally dedicated to a specific input or output type and capacity, e.g. analogue inputs 4 to 20 mA, 110 vac isolated outputs, etc. The highest capacity modules are normally non-isolated contact in or out, whereas the lowest capacity are often the analogue types. As mentioned earlier each field device should be dedicated to an input or output; additionally it is worthwhile ensuring each is fused. In some instances the modules do have each I/O fused, but if not individual fusing can save running problems and ease fault finding at a later date. When designing control circuits it is essential to observe the electrical regulations and ensure that positive isolation of switches, etc. can be achieved easily. When interfacing to switchgear it is good practice to source all control power via the switchgear isolator for each device, e.g. a motor starter isolator will also isolate the control voltage supply to the PLC I/O. The PLC power, however, should be derived from a reliable source, free from fluctuations, spikes, etc. if possible, since some units are susceptible to power supply spikes and under- or over-voltage. If units are networked particular attention should be given to the manufacturer's specification with respect to power supplies, earthing and communication cable screens. If the recommendations are not observed problems with earth loops, power supply phasing, interference, etc. may be encountered. This will sometimes apply to the power supply for the programmer when used on any unit, if an isolated power supply unit is not fitted in the programmer as standard. In Fig. 5.5 the cabinets can be supplied with the appropriate power to suit the programming unit to make servicing and programming at field level very easy.

In Fig. 5.5 the interface unit will give access for a management computer/information system of the process information required. With many PLC manufacturers this device will be available as a ready-made piece of equipment, but with others a dedicated unit will have to be acquired to carry out this duty. This will have the disadvantage common to all dedicated devices that support may be difficult and costly. Where a PLC supplier can provide video type operator terminals, etc. the facilities available may be much more than just operator controls. On many systems available today graphics, etc. are also available, which can be very useful for plant operation, training, etc. If the use of graphics is to be considered the type of graphic should be agreed by all potential users and the time to build these graphics included in the project plan, since graphics are generally purpose-drawn, some graphics taking considerable programming time to achieve a desired quality. Also the inclusion of graphics will normally give rise to additional training needs and increased project costs. Many of the larger systems houses can provide a service of pre-drawn standard symbols, etc. which can save time and costs; however, graphics costs are not easy to estimate and are normally quoted as a daily rate for programming, so if detailed outline drawings of what is required are provided, a better estimate can be given. Chapter 8 on 'Operator Interfaces, Displays and Graphics' gives much more detail on this subject.

As with all electronic equipment the installation must be carefully planned to ensure that a trouble-free system results. Observe all the supplier's specifications with respect to segregation of signal cables from power cables, etc. Plan the location of PLC and I/O cabinets to suit the plant geography and maintenance requirements and ensure communication/data highways are well segregated from all power cables and protected from possible mechanical damage. In Chapter 10 the installation of all systems is covered in more detail.

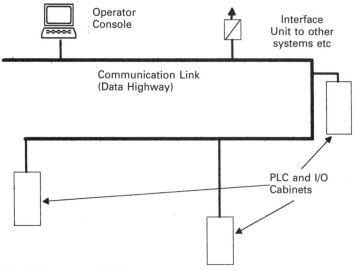

Fig. 5.5 Basic network PLC system.

5.5 Hazardous areas

Any application in an area classified as hazardous will require the applicable regulations and guidelines to be fully observed. In the case of some PLCs the necessary **barriers**, etc. can be supplied as part of the system, but for others additional equipment, approved housings, etc. will be necessary. Before inclusion of in-line items, such as barriers, ensure that the equipment is suitable to work with the selected PLC. Most PLC suppliers will be able to provide a list of suitable devices, or give a specification to be applied for their selection. If a system is to be proposed by a systems house full details of all classified areas must be provided in all enquiry documentation.

5.6   Assessing requirements and proposals

As with the small stand-alone PLC we can now generate a check list for a large or networked system to ensure we specify correctly, or confirm a proposal of, what is required. This can easily be done by answering the following questions.

1. How many control areas are required?
2. How many inputs, voltage and field device type for each area?
3. How many outputs, voltage, load of each, field device type for each area?
4. How many PLCs are needed, e.g. one per area?
5. Specification of PLC I/O, e.g. non-isolated, and total load capacity.
6. Size/capacity of each PLC required?
7. Size and type of enclosures required?
8. Maximum working temperature expected in each enclosure?
9. Specification of maximum working temperature of all proposed PLCs.
10. What operator facilities are required, e.g. how many operator terminals?
11. What communications are required?
12. What type of programmer is required and what power supply does it need?
13. Are any of the areas hazardous?
14. Is any training required?
15. What support is required of the supplier?

If your system specification includes all the answers to the above questions any good systems house should be able to assess the requirements accurately, and provide a well detailed proposal and cost estimate for the project. The above list can also be used to check each requirement in turn to ensure that a proposal is adequate to satisfy the requirements of a project. There may always be the odd item which is overlooked, or is found to be necessary at a later stage of the project, but the cost and disruption these may cause will be minimised if all the above questions are investigated and answered as fully as possible.

5.7   PLCs and continuous control

We have discussed how to identify the requirements of the process with respect to control, and in this chapter we have considered the use of PLCs. It is worth pointing out that where a PLC is used for continuous control involving **Proportional Plus Integral Action** (PI) or **Proportional Plus Integral and Derivative Action** (PID) the integral and derivative actions are time-based functions. We will not go into the mathematics of the control actions, but if, due to the loading of the PLC, the scan times change, so will the PI and PID tuning. If the scan time changes are only small this effect may not be a problem, but if the scan times change significantly some control stability problems can be incurred. It is important that tuning be carried out by a qualified technician, but no matter how well the tuning is done, it is only as good as the stability of the scan rate of the PLC.

5.8   Other systems to do the same job

If a large amount of cyclic control is required for the control of the process this will normally establish the advantage of using a PLC system. If the future development of the plant or site will encompass more continuous control than cyclic control, or if additional facilities other than control are needed, it may be worthwhile considering an alternative to PLCs. Where the control, operator display, management information, etc. requirements are high the prime alternative is a distributed control system (DCS). These are covered in Chapter 7. There are also other specialised control systems available, but these are beyond the scope of this publication, and will require very specialised knowledge to apply to a process or plant. Over recent years some specialised plant control systems have been based on mini or mainframe computers. Some of these systems are still employed, but most are becoming early victims to PLCs and DCS due to their complex, specialised and lengthy programming requirements. It may also be worthwhile considering a speciality display/interface system such as those covered in Chapter 8.

6 Continuous control

In this chapter we will assume that our investigations have identified the control requirement to be continuous control. For a continuous control application we need a device which receives information from the measurement device on the state of the process parameter it is controlling and, from this information, modulates a control device (final control device) to keep the process parameter at a desired value (**set point**).

Our flow sheet of the process should have identified where control is required, and the process operation write-up should assist the identification of where the different types of control are needed. The process design engineer should be able to identify those parts of the process where the use of special types of control, e.g. **cascade control**, can be of benefit, but if not, close study of the process operation will give a good indication of the type of control required. Some types are more obvious, e.g. 'blending' functions, which will obviously benefit from use of **ratio control** linked to flow controls similar to cascade control. For accurate blending functions there are special **blend controllers** available. In some cases special types of control can require specialist knowledge to arrange the set-up of the controller elements. If special type controls are not applied initially, they can always be considered as an extra later, if the process control stability, or functionality, is not all that it should be. For all single loops a PID controller is generally all that is required, and many will only require the P and I functions to be active to give the necessary quality of control.

All the controllers mentioned can be supplied as single items which can then be housed in a purpose-built control panel to serve an area of plant. In the case of pneumatic control this is still the norm, but with electronic controllers they can be supplied as single-loop, multi-loop or as a distributed control system (DCS). If the number of continuous control functions forms part of a large cyclic control system many of the control functions are available in PLCs as detailed in Chapter 5. The choice of PLC or DCS will depend on many factors which we will cover later when we have developed the whole picture of how the process is to operate.

Many people think that pneumatic control is no longer used, but pneumatics is still one of the most reliable power sources for operation of the final control device, e.g. pneumatically operated control valves as covered in Chapter 3. Pneumatic control is also 'intrinsically safe' and can be used in explosive environments with little or no additional equipment. Pneumatic control may have some limitations compared to its electronic contender, but it can still be of great value in some applications. The major drawback of pneumatic control is the cost of generating the compressed air at a quality to ensure reliable operation. The power required can typically be above 120 W per 100 cuft/min for 100 psig supply, and a simple calculation for the total kW hours necessary for the air consumption of the number of controllers, etc. needed for an application, will soon give the power supply costs alone. Pneumatic controllers also require a much larger degree of maintenance and regular re-calibration compared with modern electronic controllers, and this needs to be taken into account when deciding on the type of controller to be used.

If we consider the many electronic controllers available today these generally all contain a PID function (algorithm), along with high and low alarms on the measured variable input, whereas the pneumatic controller does not normally have these features. Most of the microprocessor types now available also include trend functions and limited logic capability, which can be used to turn on pumps, etc. and many can control several loops on one controller. Some of these microprocessor-based controllers are also available with facilities to be networked, allowing a control system to grow along with development of the plant. When choosing this development route one must consider equipment compatibility, since communication standards between different types and manufacturers of equipment may not be the same, and it may not be possible to form reliable links for control functions. In today's process plant 'minimum manning' is a key phrase which is always being applied, and with this in mind the expression must be applied in all areas, including operation, maintenance and installation. Single-loop controllers, or multi-loop units in single housings, require purpose-built panels which can be costly and give rise to increased maintenance along with inefficient fault finding. The cost of installation of single housing type controllers is also high in comparison to other forms of control system available today. Construction of the necessary panels and installation of single housing instruments is labour intensive; and it is not flexible for plant/process modifications or adding extra controls in a specific area if the panel is laid out in a process flow format. This does not mean that single housing instruments should not be used, but that much more consideration should be given to the ergonomics of the control panel, etc. and the panel design could be delayed until the process design is fully completed before panel design can be finalised. This may cause problems with the installation programme for the project due to delivery of essential items of control equipment.

Fig. 6.1 Single-loop controller by Camille Bauer Controls Ltd.

The controller is generally a fairly standard piece of equipment, no matter what type it is, but its performance, as mentioned before, is dependent on its being used with the correct equipment associated with it.

Fig. 6.2 Moore Products MYCRO 351 multi-loop controller.

6.1   Measurement device for continuous control

For continuous control we will require an analogue device (transmitter), giving a continuous signal output directly proportional to the measured parameter (variable). Unlike the previous application (cyclic) the analogue signal is fed directly to the controller. There are several signal standards used for the transmission of the output from the transmitter to the controller, and also from the controller to the control device, the most common being pneumatic or electronic, with live zero. The signal standards for pneumatic equipment are typically 3 to 15 psig (20 to 100 kPa), and for electronic are typically 4 to 20 mA. Today we have another family of signal standards under development in the form of digital communication. Until there is an accepted international standard there will be several signal standards to consider. An early example of the new standards is the signal used for communication to smart transmitters. Various manufacturers have developed signal standards to suit their equipment, e.g. HART™ by Rosemount (now Fisher-Rosemount). HART™ is now available on many pieces of equipment other than those manufactured by Rosemount, and has become one of the new signal standards. To maintain equipment utilisation the digital signal is superimposed onto the 4 to 20 mA dc signal. This allows the transmitter to be used in conventional control loops as well as control systems capable of accepting the digital signal directly. The use of a separate programming/configuration device allows calibration of the unit without the need to remove the unit from the field or stop the process. With the advent of FieldBus, and other similar standards, the digital signal standards are being further developed to allow networking/multi-drop connection of devices in the field, including transmitters and control valves, etc. Ultimately the intention is to allow all

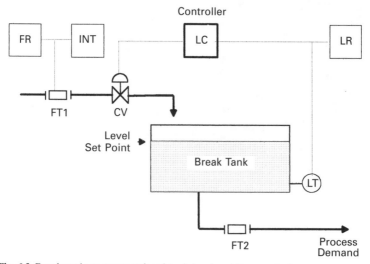

Fig. 6.3 Break tank or constant head tank level and flow control.

field devices to be networked/multi-drop into the control system, saving cable costs, etc. and giving calibration, status information and local control functions of field equipment at control terminals, resulting in reduced maintenance and engineering costs. The selection of the measurement device required was covered in Chapter 2. Filling in as much detail as possible on the transmitter specification sheet at an early stage will allow other functions in the project to continue such as the choice of the control facilities required which will lead to the selection of the controller.

6.2 Control device for continuous control

For continuous control the control device must be capable of being modulated continuously to achieve the degree of control required. Some of the regularly used devices include control valves, variable speed devices (e.g. pumps), actuators, etc. which we have covered in Chapter 3, but here we will identify their role in continuous control.

At the stage of process design many of the control devices will have been identified. The quality of control, however, may be improved by closer study of the device and its suitability to the application. The control response must be of comparable speed to the process speed, i.e. 'if we have a fast process a fast control response is required'. For the fast process a fast responding control device is therefore essential, whereas if we have a slow response requirement we can always slow down the action of the control device. *But we cannot speed it up beyond its design limitations.*

Many unstable control applications are the result of the wrong choice of control device or bad process design (e.g. tanks, vessels or pipes too small relative to the throughput). A common fault in process design is 'designing for the maximum'. This is okay if the plant is to be run continuously at maximum throughput, but quite often the normal operating rate can be well below design maximum and may vary considerably. This can result in over-sized equipment which is costly to run as well as difficult to control. As we have seen previously a guide to the ideal point of control is for normal throughput to be approximately 50 to 60% of control equipment range, including measurement device, controller, and control device. If a plant is to be run at maximum design throughput the running control point should not be more than 80% (90% absolute maximum) of equipment range. If there is no capability for the controller to regain control of a process parameter following an increase in demand there will be a loss of control, which in the worst case could lead to total instability.

Let us consider a simple process with all the equipment necessary to perform the control and information recording for the process. If in the example of Fig. 6.3 the flow through the flow transmitter FT2 exceeds the capacity of control valve CV, the level control LC will not be able to maintain the level at set point until the flow through FT2 falls to within the capacity of control valve CV. The time to regain level set point will depend on the difference between the capacity of control valve CV and the flow through FT2. In this example the control valve CV is undersized if the control is to be maintained at the high set point in the above conditions. In this example undersizing of control valve CV has created a situation where the controller cannot maintain control in all situations. Also if the characteristic of the control valve CV is a conventional butterfly valve and the controller is tuned at conditions of maximum opening of the valve (control signal response at minimum) to maintain set point the control is likely to become unstable when the flow through FT2 becomes low and the controller regains set point, since the valve CV will then be open a small amount (response to control signal change is at maximum). This situation can quite often result in the valve CV 'bouncing' on its seat giving open/shut pulses to maintain level at set point. In many applications this situation can be tolerated, but it could lead to premature failure of the control valve if it persisted for a prolonged period. If the facility is available on the controller LC, a logic over-ride can be used to keep the control valve CV closed when flow FT2 is zero and level is at set point. As can be seen the sizing of all the items in the loop is within range except the control valve CV. This is only one example out of many where the control device is critical to the good operation of a control loop. The type and performance of the control device is critical for good control, just as much as good process design. Do not use a control device because it is already there! Ensure it is really suitable for the duty in hand.

With most control problems there are generally three main causes.

1.   Control device sizing/suitability.
2.   Process design (e.g. too much turn-down for the control loop to manage, or poor positioning of instruments).
3.   Poor controller tuning.

In the example in Fig. 6.3 we have seen condition 1 caused by the level control valve CV being undersized. A similar condition could be generated if the pressure drop across flow transmitter FT1 causes a restriction in flow prior to the control valve CV, limiting the control valve maximum flow. Condition 2 is demonstrated if the process demand has too much turn-down, the pipe size to control valve CV is too small, or the break tank is not of sufficient capacity to accommodate the process demand variations. Condition 3 will be covered in Chapter 11 which deals with commissioning and troubleshooting. There are other factors which will cause instability, but these are generally the result of complex factors which would require the skills of a trained Instrument/Process Control Engineer to correct. With modern controllers employing microprocessors facilities such as auto tuning, etc. are available. These facilities are of great use when commissioning, but should not be relied upon to overcome basic design or sizing problems. The recorders in Fig. 6.3 can be used to identify the problems of sizing, etc. very easily. If the flow signal from flow transmitter FT2 plateaus at its highest value, and the level continues to fall, the supply does not satisfy the demand. If the control is always regained before the break tank level falls too low, the break tank size is adequate and the problem is one of sizing, or restriction, of the outlet devices, and beyond the capability of the controller, i.e. condition 1 or 2. In many situations this state of affairs is the prime reason for the break tank being installed, and no further action is necessary. The controller can be set to give the quickest recovery rate possible and, if available, alarms on the controller can be used to identify conditions of critical low level, etc. to advise the operator of potential problems. If the break tank in Fig. 6.3 is now used as a constant head tank, the fluctuation in level must not occur, since this will negate the constant head condition. In this instance the control valve CV, or the pipe size to the valve, must be increased to ensure that the maximum valve open condition does not occur. The break tank service could be satisfied by on/off control, or a ball-float valve in place of CV, but for constant head applications the continuous controller, level transmitter and control valve loop is necessary to satisfy the control requirement of precise level control for the process. If the actual 'head' generated is critical to the process the recorder LR should be included for process fault finding and alarms used, set to a tolerable dead band so as not generate unnecessary alarms.

6.3   Controller to suit the application

After careful study of the process design we should now be able to distinguish where the different control types should be applied, and some of the additional facilities that will give the necessary operator and management information required. If we employ single-loop controllers, etc. we will need to house them. If there are only a few loops to be considered then perhaps a purpose-built control panel will suffice, or even a multi-loop controller housed in a simple cabinet. When using multi-loop controllers each unit must be carefully considered since the controller will normally display only one control loop at a time. It will therefore be necessary to identify all those controllers on loops requiring regular operator changes, or observation, and those that will need only occasional monitoring/access. The controls needing constant monitoring/access should each reside on an independent controller, and the low priority controls can be accommodated on the 'background' controllers of the multi-loop units. This may result in a higher count of units than initially estimated, but may be necessary to ensure adequate, and safe control of the plant. Another factor often overlooked is 'how many controls will need constant access during start-up', since there may be additional requirements at this stage. This might add further to the count of units required, by changing a 'background' control to higher status on a multi-loop controller where the primary control is still requiring constant access during plant start-up conditions. When a large number of control loops are involved the purpose-built panel becomes very large and expensive. Also it may take more than one operator to cope with the size of panel which will be needed. The method of operation, ergonomics of panel layout, number of operators available, management and operator information required will add to the factors affecting the decision on the type of controllers, etc. chosen. The answer to a situation of requirements for large panels, extra operators and lack of space for control interface equipment and operating personnel can be to use a DCS.

Fig. 6.4 Selection of Fuji Microjet Recorders available from Coulton Instrumentation Ltd.

7 Distributed control systems (DCS)

When a large number of continuous controls, recordings, alarms, special control functions, etc. are required the DCS can be one answer. As the title suggests the system is designed to carry out all necessary control functions for a plant/process over a network of controllers distributed over the plant and linked together with operator terminals, etc. The first successful DCS was the Honeywell Controls TDC™ system (Totally Distributed Control). Today this has been upgraded to become the TDC 3000™, with the original concept and many of the original techniques still employed. The TDC™ system has become the benchmark for many of the competing DCS designs now available, and is still regarded as an industry standard for many processes. Other DCS manufacturers include Fisher-Rosemount (System 3™, DeltaV™ and Provox™), Foxboro (HIA™ and Spectrum™), ABB (MOD 3000™), Moore Products (APACS™), plus many other manufacturers and models. All DCS architecture is generally of a similar layout with operator consoles linked to control cubicles housing the controllers, etc. via a powerful, fast, high integrity communications system. With all DCS it is the powerful and secure communication system which allows the 'distribution' of control. The predecessors to DCS were multi-loop type, mini computer-based units, where the I/O capacity could be increased by adding memory, but the total size was limited to the processor/maximum memory size. Some of these systems are still working today, but are now gradually being replaced by DCS as component availability becomes a problem, due to advances in computer/electronic technology. The parallel development to DCS was the multi-loop controller mentioned earlier. If one considers the power of the minicomputer-type unit combined with the flexibility of the multi-loop controller this closely describes a DCS. As advance in technology has taken place the DCS has rapidly expanded its capabilities. All systems available today can perform very advanced control functions, along with powerful recording, totalising/integration, mathematical and some decision-making functions. Many can be tailored to carry out special functions which can be designed by the user and many will now communicate to, and integrate into, management computing and information technology systems.

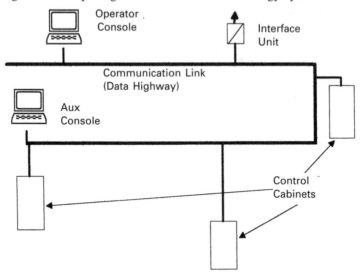

Fig. 7.1 Basic DCS.

Figure 7.1 shows a diagram of a basic simple DCS. The interface unit is the means by which the DCS transfers information to the management computer or information system; it converts the communication protocol used for the DCS to that used for the management system. In many systems this is a two-way device allowing production information to be transferred to the management system and to receive information from the management system, e.g. production schedules etc. The controllers are housed in the control cubicles and these are normally located close to, or in the field, where the control is required. The communications highway/ data link (different names according to manufacturer) is normally in the form of a dual communications network, where both are active but each is capable of carrying all the communication should the other fail, any failure automatically switching all the communications over to the good system. Different manufacturers have differing communication standards, but hopefully in the near future all systems will be common. There are active committees, such as Fieldbus and World-FIP, working to achieve such a standard within the next few years. It is envisaged that all existing participating manufacturer's equipment will not be rendered obsolete as a result of this. Most of the existing DCS will be fully supported, and with minor modification will be compatible with the new standards. The basic system shown in Fig. 7.1 can be expanded to handle many control loops and functions. A typical system size can easily be over 1000 loops, but the make-up of the different functions required is the governing factor on the system the capacity of a single communication system DCS be insufficient, or the communication highway be of insufficient length, a second or third as required can be linked to the original network to achieve the requirements. As a guide to the size of each network the maximum capacity of a central control room for an area of plant, manned by several operators, can easily cover 8 to 10 local plant control areas, depending on the level of information to be displayed for any area. Each local plant area can be of many continuous control loops, maybe over 100, along with motors and other equipment, again possibly in excess of 100, the restricting factor being how many critical controls there are. There should be no more than the number one operator can handle at start-up or in an emergency condition.

7.1   Integrated control

Distributed control systems are not limited to continuous control applications; they are also capable of carrying out many of the functions of a PLC. The inbuilt design function, however, is of *integrated control* where all the general plant operation functions of control, whatever the type, are carried out by the DCS. These functions include continuous control, cyclic control, *logic control, motor control, batch control*, etc. The difference is in the processing speed of sequence functions. Where a PLC scan time is in milliseconds, a DCS will be in seconds, or part seconds (typical fastest speed 0.25 sec). In most process control applications the speed of the DCS is quite adequate, but if high speeds are necessary in selected operations PLCs can be interfaced directly to most DCS using standard components which make the PLC transparent to the operator. On some DCS this problem has been addressed with the use of high-speed controller cards, which are part of the system and give the very high speeds necessary for special applications. These controllers come with a cost premium, however, although they are cost-effective where only a limited number of high-speed functions are necessary. The standard functions available with DCS today provide an integrated control system capable of all aspects of processes control, able to be applied to most process, either as the system stands, or with integration with other equipment, to achieve the total control requirements of any plant, and in many cases the necessary management production information.

7.2   Batch control

The DCS is also finding increasing popularity for *batch control*, particularly in pharmaceutical applications. Batch control is a control application where numerous recipes need to be applied, full documentation of the production batch is required and a very easy system of recipe change / modification must be possible. The nature of some DCS allows build-up of the recipes required in a modular format, which saves re-programming/configuring the system for each change of production batch/recipe. With increased quality standards now required, all manufacturers are needing more production information and documentation, so the control system is the first choice for obtaining this information. Many DCS vendors provide systems with pre-configured software, allowing very quick application to any process, with little special software to be written for most applications.

7.3   Configuration

The terminology used to describe how a DCS is programmed and applied to a process is commonly called *configuration*. Most DCS are provided with the software pre-programmed, so configuration is required only to tailor the DCS to the application. This overcomes the need for any special software language to programme the microprocessors used in the system. All this level of programming is already carried out by the DCS manufacturer. This enables a DCS to be applied to any plant very quickly, and cuts down the de-bugging necessary on programmed software. Many DCS can also be re-configured without need to take the system off-line which enables plant modification, etc. to be carried out in the minimum of down-time, providing the necessary precautions are taken to ensure integrity of the running process. This facility can be of particular use where continuous type process, with short periods of shut-down maintenance, are to be modified. Some systems, however, do have a facility for the software programming of parts of the system in special circumstances, to enable the DCS to cope with special plant control requirements, etc.

7.4   Graphics

All DCS available today have graphics for operator terminals, etc. Like the graphics available for PLC systems, the time necessary to draw graphics can be very considerable, and needs to be included in the project plan. Again a good selection of drawings of what is required will reduce graphics drawing time, costs, etc. Many of the DCS manufacturers can provide some pre-drawn graphic symbols, and even some of the more common process plant, e.g. boilers, heat exchangers, etc., which can save time and cost. Unfortunately in most cases graphics are plant-specific and hence time-consuming to produce, especially when control is via a fully functional interactive graphic. With most DCS graphic drawing requires special training to achieve the best results, so consideration should be given to the supplier doing this work if the number of graphic screens required is fairly small. Graphics are covered in more detail in Chapter 8.

7.5   New communication standards

Earlier there was mention of standardising communication on DCS, PLCs and ancillary equipment. These standards will potentially change the architecture of DCS and PLC control by allowing the introduction of **Local Area Networks** (LANs) for connection/communication of field equipment, including control devices, measurement devices, etc. giving 'multi-drop' communication serving many field devices. The aim is to establish a standard as universal as the 4 to 20 mA signal standard used today. Additional to this will be other communication standards serving the DCS highways, and the communication with management computing and information technology systems. Since this communication development programme is well under way most of the DCS and PLC suppliers are designing today's equipment with facilities to accommodate the standards as they become available. The implementation of the first stages of Fieldbus and World-FIP should be well in place soon and hopefully the major standards should be completed early in the next century. As a result of this the architecture of DCS and PLC systems will start to change over the next five to eight years, giving more facilities and enhancements of existing functionality. The aim is also to allow mix of manufacturer's equipment on the same communication highways, etc.

7.6   Working environment

Distributed control systems are available for all types of working environments likely to be encountered by the system. In most cases this will be accomplished by use of suitable cabinetry, housings, electronics, etc. In the case of hazardous areas additional equipment is often required, which should be made part of the system suppliers' responsibility if at all possible. This will enable the appropriate modifications, etc. necessary to be integrated with the rest of the DCS so minimising site installation work and allowing off-site testing and approval to be completed before delivery. Where special cabinetry is proposed by a DCS supplier examine the proposal closely to ensure additional services are not to be provided, e.g. forced air cooling, etc. Such services can add considerable cost to a project and may require large holes in walls or floors, which will involve extra civil work to be arranged.

7.7   Site controls development

The DCS is possibly the most powerful and flexible tool available today for control of any process or plant. They come with a host of facilities which require little addition to satisfy the majority of applications. For this reason alone they can be applied with confidence that any extra requirements, identified at any time, can be added with minimum disruption. Generally only the costs of I/O, transmitters, control devices and installation, etc., plus the necessary configuration, will have to be covered. If a DCS is installed it is very worthwhile ensuring that the latest software level is applied, at least once a year, to be able to make use of on-going development of the system and avoid significant re-training of configuration and maintenance personnel when major up-grades and expansions are carried out. If a minimum maintenance/support contract is taken out with the supplier to include the software up-grades a cost-effective back-up to site maintenance can be achieved. Such contracts will give the opportunity to plan up-grade expenditure, etc. over many years and allow better maintenance/repair budgeting. Hopefully such contracts will ensure that supplier personnel familiar with your DCS layout, plant and process will be the regular contact whenever problems arise and visits to site are necessary. This familiarity will pay good dividends should problems occur. The penalty for not keeping a system up-to-date is that when an up-grade is undertaken the 'on-cost' for bringing the system up to a standard to allow the addition of new équipment can be very high, and possibly too much for the project to carry! Development of the DCS will also benefit by the support having a good knowledge of the site and system, as they will be able to advise when software up-grades will allow additional facilities to be utilised to achieve improvements in control, reporting, etc., while requiring site personnel to do only a small amount of additional configuration.

7.8   Specifying what is required

We should now be able to generate a comprehensive list of questions to be answered to ensure we specify correctly, or confirm a proposal is accurate for what is required.

1.   How many plant process areas?
2.   How many areas are hazardous?
3.   How many operator terminals required?
4.   How many analogue inputs and field transmitter types per area?
5.   How many analogue outputs and control device types per area?
6.   How many contact I/O (or how many motors, etc.) per area?
7.   What level of spare analogue and contact I/O capacity should be accommodated in the proposal?
8.   How many interfaces to other systems are required?
9.   What communications types are required to interface with other systems?
10. How many graphic screens are required and are these to be interactive?
11. Who will do the configuration?

12.   Who will draw the graphics?
13.   Who/how many will require training and to what level?
14.   What level of on-going support will be required and for how long?

By answering, and providing the answers to, all the questions above, a good DCS supplier will be able to understand the project requirements and generate a comprehensive proposal for a DCS to suit the application, without the need for you to understand the architecture of the supplier's system. The supplier should also be able to generate a function specification of how the proposed equipment will satisfy the requirements of the process, plant operation, management information, etc. from the answers to the above questions. If this information is not provided in full, the prospective supplier may not properly understand your requirements and many have a problem supplying what you really need to do the job correctly. The customer should not be expected to produce anything more than the requirements specified in his or her native tongue, and should be provided with a response from the supplier of the functional specification, and full details of the components of the proposed system, in the same format. Failure to fulfil this requirement should preclude that supplier from selection. Additionally the supplier should be able to provide an installation specification immediately an order is placed, or even at the proposal stage, to allow the system installation to be integrated into the project plan, and allow installation contractors to be able to quote for the DCS installation once cable schedules, etc. are available, if the DCS supplier is not involved in a turn-key project. Any proposal should be systematically checked against the project 'specification of what is required' prior to any order placement to ensure that the proposed equipment satisfies *all* the project requirements. With most DCS any items overlooked or added at a later date can be easily accommodated, but good specification of requirements initially will minimise overspends, shortfalls, etc. and of course embarrassment!

7.9   Future DCS functionality

Many of the DCS suppliers are building advanced functionality into their systems. Some suppliers already offer 'Fuzzy Logic' control and are offering other equally advanced facilities in the near future such as 'Neural Networks', etc. All major suppliers recognise the need to communicate with other equipment and are working with the major communication standards such as Fieldbus and WorldFIP mentioned earlier. These facilities and functionality will give any user a system with up-grade potential and expandability undreamed of a few years ago. With such on-going developments by the manufacturers the user must make allowances for continuous up-grade, possibly only software, of the installed equipment to take advantage of the progress. Much of these up-grades will inevitably require the

Fig. 7.2 Moore Products APACS™ DCS operator terminals.

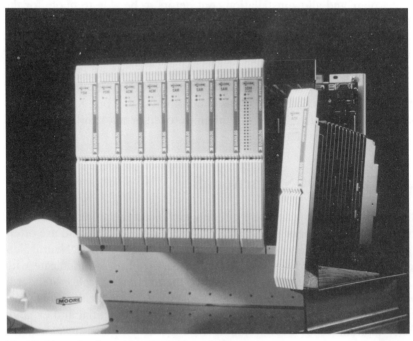

Fig. 7.3 Moore Products APACS™ DCS controller cards.

equipment to be off-line, necessitating the plant to be shut down, so an annual plant maintenance shut-down period is possibly the most convenient. This situation of being able to keep up with the latest functionality, etc. is currently only available to DCS users. In the case of PLCs there is at present a requirement to change out much of the hardware to keep the same level of advancement. Obviously some DCS up-grades require hardware changes, but many software up-grades can be done before major hardware up-grades become necessary. Most system suppliers will be able to give details of software up-grade paths for their equipment, and should be able to offer a support package for such improvements. Such support packages can also include periodic maintenance/checkout, breakdown support, etc. by the supplier, and the costs can be built into the on-going budget for the system giving much closer control of the cost of ownership of the control system.

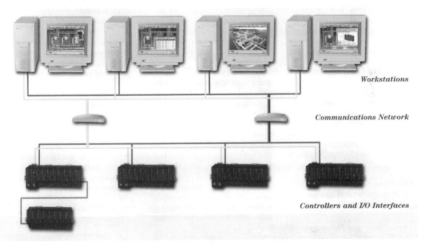

Fig. 7.4 Architecture of Fisher-Rosemount DeltaV™ DCS (courtesy of Fisher-Rosemount).

7.10 Intelligent motor control centres

With the advancement in the use of computer control has come the application of the technology to motor control. Today there are an increasing number of manufacturers of **multi-motor control panels** (MMP), or **motor control centres** (MCC) offering control interfacing to PLCs and DCS enabling the control of the electric motors connected to the intelligent MMP or MCC over a communications link. This development gives a further saving in installation costs and makes more information available to the operator and maintenance engineer. Utilising the communications link saves costly control cabling and, in the event of plant and equipment damage, such as that caused by small fires, the replacement of communications cables is much quicker and easier than that of dozens of control cables, enabling the plant to be brought back into production much quicker, and with less cost. We cover the installation standards for signals and commu-

Fig. 7.5 Drawing of the smallest step of a Fisher-Rosemount DeltaV™ DCS (courtesy of Fisher-Rosemount).

nications in Chapter 10, but it is worth emphasising here that the quality of control is directly proportional to the quality of the installation. Most important, however, is the need for communications to be completely free of interference. Attention paid to the communications cable installation will pay dividends in control reliability many times over.

7.11   Systems development

Like all applications involving microprocessors, the on-going development of the DCS capabilities is no exception. HART™ has already been mentioned, and with some DCS this has been encompassed into the configuration capabilities so field devices can be ranged, etc. directly from the DCS engineering terminal. With the increasing availability of more powerful software, such as new Microsoft-Windows NT™ packages, some suppliers are encompassing this into their product to give enhanced facilities as well as other benefits. Training for configuration/programming is one area where major advances have been made with 'on-line reference text', along with exacting guidelines, enabling any engineer to learn quickly how to configure the system for basic and many complex tasks, without the need for lengthy, and costly, training or reference to costly, bulky manuals. Being able to learn only what is needed for the project requirements reduces the training time and enables configuration to be done in a very short space of time. Typically a five to ten fold reduction in configuration time compared to PLC programming for control of the same process can be achieved with this type of software. Additionally the on-line help is used to warn of wrong configuration when it is tried, plus off-line simulation reduces the level of de-bugging necessary. Primarily all DCS can accommodate on-line card changing, as well as configuration change, but now simulation can be run while the system is operational. This is particularly useful where recipes need developing, which can now be done without shutting down the process. The use of powerful software, which is similar to other regular applications, means that there is little mystique with the system, and also offers a powerful means for both system and process fault diagnostics. Additionally the capability to document the complete system set-up, and the configuration, enables a significant reduction in providing as-installed documentation. With such developments it is essential that a careful study of all available control systems is carried out to ensure that the best suited and most cost-effective control system for the project is selected.

8 Operator interfaces, displays and graphics

Information presentation to the operator is of prime importance for the success of the process control aspect of a project, as is good acceptance by the operating personnel. Complicated or confusing presentation of any information will invariably lead to rejection by operators. Displays take several different forms and will depend on the type of process control equipment employed. Over recent years various video type displays have become more readily available, but there are many other types of information presentations which may be more suitable to the application in hand. For most small control systems a fixed format display is probably the most economic, but with current communication technology most controllers with fixed displays can be interfaced to sophisticated displays if required. The type of display is generally no longer limited, and careful choice of control equipment can give all the process information required, providing there is adequate measurement equipment, etc. installed. Some control systems can have closed circuit television camera pictures shared on the display screens, which can give an even more enhanced level of plant control.

8.1 Single-loop controllers

Most single-loop 2 or 3 term controllers have fixed display formats showing set point, measured variable and output, along with other status information such as whether the control is manual, local automatic or remote. Obviously the display format cannot be changed on the controllers themselves, but the arrangement of the controllers in control panels can, and this is of great importance. The ergonomics of the panel layout is especially important when a large amount of equipment is mounted in it. The controllers should be mounted at eye level, and there should be easy access for operators to make changes. Where several single-loop controllers are to be mounted in one panel, care should be exercised as to their function. Generally follow the process as to the positioning of the controllers, i.e. follow the process flow from left to right. Some modern single-loop controllers have facilities to allow communication via one, or other, digital communication system. If equipment is selected having this feature the data on each controller can be transmitted to a data-logger, management computer, SCADA unit, etc. With any purpose-built panel the build cost will be high, and any changes to the layout after commissioning will invariably look obvious and be difficult to carry out if involving several items.

8.2 Multi-loop controllers

Having first ensured that the prime loop on each controller is the default display, follow a similar arrangement to that for single-loop controllers detailed above. Additionally ensure that the secondary loops on each controller are truly secondary and follow a similar logic of location on the controllers and panel position, even though they will normally not be the default display of the controllers. Like the single-loop controllers some modern multi-loop controllers have facilities to allow digital communication to other equipment. If controllers are selected having this feature, the data on each controller can again be transmitted to a data-logger, management computer, SCADA unit, etc. The limitations and problems associated with single-loop controllers mounted in dedicated, purpose-built panels will apply equally.

8.3 DCS displays

With any DCS the ergonomics are generally pre-determined, and the displays and keyboards should be carefully studied to ensure that those most suitable for the project are selected. Presenting an operator with a QWERTY keyboard when he or she is not familiar with one, and also when it is unnecessary, will cause needless concern. Most DCS loop display faceplates are pre-formatted and only require arranging into the logical groups, within the group display limits, to suit the process. Again the arrangement of the groups should follow the format as described in the arrangement of the single-loop controllers in a panel, i.e. following the process flow. The great advantage of DCS displays is that any loop faceplate can be repeated, as required, in as many group displays as necessary. In some applications one or more loops may have interaction with other parts of the process. If this is the case, the loop faceplate can be repeated in all the groups where it has influence. Additionally modern DCS have graphics facilities which can normally be fully interactive. On some systems graphics building is complex, and hence specialist programmers will be needed to interpret the plant requirements into suitable displays. Most, however, have graphics building which is very simplified and well within the capabilities of whoever is trained to configure the system with minimal additional training. Many symbols are pre-drawn and only require copying and placing on the appropriate display page, and with the addition of lines, additional symbols, etc., will form a very adequate graphic display for most plants. Most of these graphics can be made interactive by linking various parts of the display to the appropriate loop faceplates. The result is a powerful display through which the plant can be controlled as well as viewed. The action of linking many of the symbols will give animation of the display, such as tanks filling, pumps changing colour when running, etc. plus display of the running values of the process.

On most DCS there is a facility to display overviews of groups which can give a plant status monitoring facility covering dozens of loops and motor controls on one display. The majority of DCS will also give fully interactive displays and controls of PLCs interfaced to it without the need of a separate SCADA terminal. Another facility available is the log function. Generally there are two types: alarm, and event. The size is variable to suit the application and the broadcast and recording can be set up to suit the plant control. Finally there is also the function of reports giving production and consumption details, etc. Mostly the alarm and event reporting is pre-formatted, but the other reports can be set up to suit the requirements of the plant on the majority of DCS.

From the above comments you may possibly get the impression that a DCS cannot interface with a SCADA package. This is definitely not the case; but why employ a separate display package when the DCS may have more than adequate facilities to meet the needs of the plant. Additionally most DCS come complete with all the facilities covered in the software licence, but a SCADA display will incur an additional software licence. Most DCS have standard digital communications available which will easily allow interface to SCADA and other packages. Consequently if it is necessary to interface with existing SCADA systems, etc. this can easily be accommodated. With older, slower systems, however, the update and response speeds may be slower than desirable, and this has to be considered from all aspects, especially safety of plant and personnel; e.g. what is the acceptable delay on alarm display and response? It is not advisable to have an alarm klaxon, activated by the DCS or PLC, sounding several seconds before the alarm is actually displayed on the SCADA display screen! You may think this is an exaggeration but I have experienced just such a situation!

8.4   SCADA and software licences

Supervisory Control and Data Acquisition (SCADA) is the term given to the operator interface to a PLC network, networked intelligent controllers, etc. All SCADA systems generally employ PCs loaded with special software, programmed to customise it to suit the control medium and the process under control. It is the software that turns the PC into an interface with the controllers for operators and management to obtain information from, and give instruction to, the controllers, be they PLCs, communicating single or multi-loop controllers, etc. No matter what type of control is used the SCADA system will require a licence for its use. Sometimes there may be more than one licence needed. A 'run' licence will always be required, but normally a graphics build, as well as communication licences, will be needed to operate the SCADA package to cover all the site requirements. Remember such licences are applicable to each point of use, so a 'run' licence per terminal will be required, and possibly only one licence for the graphics build, and a communications licence to cover just the number of taps on the communications applicable to the SCADA information and data transmission. Where a SCADA package of software uses an operating system, such as Windows NT™, a licence for this will also be required for each terminal using the operating system software. These licences are a legal requirement under the Software Protection Act, and can add a considerable cost to any project. When costing a project it is advisable to ensure that all the necessary software licences are budgeted for. Whoever is invited to quote for the system should be able to include all the licence costs you require, but remember you must ask for them to be included. A point to note: purchase of a software licence is not a guarantee that the software will work on your system or work with other software. Additionally on-going update/revisions of the software are generally not included in the licence fee. With most software registration will give a lower cost up-date/revision path as well as access to help desks, etc.

8.5 Disk drives

It is worth mentioning here that when using standard PCs the disk drives are very vulnerable to ingress of dirt and moisture. Whenever a PC is to be used for operating a plant it is very worthwhile considering using 'hardened PCs' with remote disk drives housed away from the poor environment. A control room may be air conditioned, etc., but if operators are likely to be operating the system with dirty, wet hands the standard PC will be vulnerable, especially if they are possibly handling disks. Also keeping disk drives away from operators, and accessed only by authorised personnel, will minimise the risk of contracting any PC viruses!

8.6 PLC displays

With PLC systems the ability to display any information to operators and management will require the addition of some form of display equipment. Most PLCs will display the status of the inputs and outputs at the field interface modules and also on a programming unit, if connected, but these displays are for maintenance and programming and not for general operation of the PLC. In some circumstances a start input, along with running, fault and cycle completed indication, may be all that is required, and this can be satisfied by simple indicator lamps. In the majority of applications, however, more information is necessary for plant control and operation, which will require more indication lamps and push buttons, or a more detailed display of some sort. There are many types of display unit, ranging from single-line types, to multi-page, microprocessor driven video displays. All types have particular benefits, and it is essential to establish all the plant operation requirements at an early stage of the project before deciding on any particular type of display unit/operator terminal. Many of the display units require special programming techniques, especially where dynamic, interactive graphics are wanted. Additionally if several PLCs are required to interface through one display terminal these will generally have to be networked to achieve the required level of control, and display. If there is more than one type of PLC to be networked, even from the same manufacturer, this could limit, or even render impossible, full communication through one display/operating terminal. For this reason alone it is essential to have a good feel for the final scope of the control/display requirements of the project. If there is any doubt on sizing, one should always err on the side of the higher capacity system to avoid future critical limitations. As a guide, if any interface system has a fixed maximum capacity this should always be estimated to be at least 150% of estimated maximum requirements for the project. Once a system is commissioned, the operators and management will want extra information and controls which will soon swallow up the spare capacity, unless for any reason the control and displays are rigidly fixed at the time of commissioning.

8.7　PLC (or DCS) and SCADA performance guarantees

The PLC program is quite often the responsibility of one group of individuals, and then the display programming is the responsibility of others. *This can more often than not be a recipe for disaster!* For a successful integration of PLC programme and display/operator interface, there should ideally be only one source for the total integrated package. Sub-contracted or separate departments should not be entertained if at all possible. In any event, only one supplier should be charged with the overall responsibility for system performance. At the early stages of any project there can be the appearance of much goodwill amongst and seemingly convincing promises from a group of suppliers, but when things go wrong the blame is invariably shunted from one party to another if more than one source of supply is involved. If the client is not fully familiar with the equipment and all aspects of the programming, etc. it can be very difficult to identify the source of the problems, and hence the responsibility. With more than one supplier the small print of the supplier's terms and conditions, when involved in joint ventures, can often form the means by which they can absolve themselves of responsibility, or at best render extra charges for additional time, etc. to sort out the problem. On the brighter side many PLC suppliers are able to provide complete packages and numerous systems houses can provide equipment, programming, commissioning and training services, along with a performance guarantee to satisfy the project requirements. No matter how small the project, if PLC and display/operator terminal programming is not to be undertaken in-house a supplier or systems house capable of supplying all the required programming and support is the safest route for a successful conclusion to a project.

When integration of several PLCs is required it is of vital importance to generate the overall plant control philosophy detailing all the operator and management information requirements very early in the project. The systems house, no matter how good they are, do not know what you really want, they only think they do! Like specifying all the applicable data to obtain the correct measurement, or final control device, it is equally important to specify all the requirements of the completed control system. Many manufacturers of process equipment will provide a pre-programmed PLC to operate their equipment correctly. Linking these pre-programmed PLCs together on a network will *not* normally give good overall plant control. The complete plant operation should be carefully defined, including how different pieces of equipment need to interface and interact to *achieve the requirements of the process, not the requirements of the pieces of equipment*. Many process plants have been designed, installed and commissioned only to find that the expected process functionality doesn't exist. It is worth considering the people required to work together as a team to design, build and operate a process plant. No one person can work as an individual; similarly no one piece of equipment works completely individually; it is part of an integrated process, and as such its required operation must be considered when generating the plant operating philosophy. By taking this into account it should be possible to identify the changes necessary in the PLC programme to achieve the functionality required. Operator information must also be considered. It is all very well protecting a piece of equipment using an interlock, but the operator needs to be informed of each interlock preventing the piece of equipment from functioning correctly. Many hours of plant availability can be lost by poor operator diagnostics information of process as well as equipment problems. Generating the process plant philosophy may take several dozen man-hours, but can save hundreds of process plant availability hours! Often the hourly rate of plant availability is hundreds of times greater than the cost of the man-hours to generate the original operational specification.

8.8   DCS programming/configuration of operator terminals and graphics

DCS programming/configuration generally does not have the same pitfalls as PLCs integrated with SCADA packages. Due to the basic design, the equipment is normally of common supply and communication standard from any one supplier. The supplier is normally able to supply a fully programmed system, with performance guarantees, along with all levels of training, to enable the end user to carry out all the levels of support/programming suited to the site personnel. PLCs can be interfaced to many DCS utilising the DCS operator interface. In some cases the display may be of minimal PLC status, but in others full integration may be possible. When integrating a PLC with any DCS, the scope of supply, and hence responsibility for performance, is still an issue, and should be fully understood by all parties before any orders are placed. The chance of major problems, however, can normally be assessed at the tendering stage, provided the full scope of the project control and display requirements are identified to both the DCS and PLC suppliers at the enquiry stage. Most DCS suppliers will take on the performance guarantees if they interface to a PLC approved by them, especially if they are made responsible for the total control supply. In such cases the DCS supplier will generally offer a full scope of equipment, programming/configuration, training and commissioning of a control system, including group displays, interactive graphics, report configuration, etc. Where individual pieces of equipment are employed using their own dedicated PLC for protection, etc. the information can be transmitted to the DCS in the same way as it is transmitted to a SCADA system. The diagnostic information can generally be applied using a standard format of displays on the DCS terminals, with the result that all interlock information is displayed in a common manner for all pieces of equipment, so easing operator understanding. The information has to be identified, however, and configured into the relevant displays, so there is an equal necessity to generate a process operation philosophy outlined above.

Graphic generation on DCS generally uses pre-drawn symbols which are configured into the graphic, and linked to the appropriate device parameters within the system, to make the graphic dynamic. With SCADA graphics, on the other hand, the often pre-drawn symbols are programmed into the graphic display in a similar way to using a computer-aided design (CAD) system. This generally requires a higher level of skill to achieve a presentable graphic display. The reason for this difference is that the graphic for DCS is dedicated to that system and the communications within it. The SCADA systems are generally PC-based, and tend to use Windows-based software to make the SCADA as universal as possible. The communications required to make the graphics dynamic should also be universal to ensure that the display can be interfaced to as many PLC types as possible. Each DCS, and PLC/SCADA system graphics have their advantages, or limitations, but for ease of generating basic control graphics the DCS offers a package that will normally work without any communications difficulties. On some SCADA packages there may be a need to obtain a licence for the software to generate the graphic initially. Before placing an order for equipment, ensure that all the functions you want carried out are covered by the licences you buy with the system.

8.9   Graphic colours

Many graphics are very good but look poor because of the colour scheme used. Operators have to use the graphic for many hours and the colours will significantly influence their acceptance of the displays. There are guidelines available for many applications which any systems house, DCS or SCADA software supplier will be able to supply. Vivid colours may draw attention, but if the state generating that colour, such as an alarm, persists unnecessarily it will soon become an annoyance, and may then be ignored to the detriment of the plant operation. Flashing symbols, etc. can also annoy after prolonged periods. There is nothing wrong with the use of vivid or flashing displays to indicate alarm status, etc., provided the acknowledgement of the state by the operator reverts the display to an acceptable colour/state of display. If the situation persists then reinstate the flashing, etc. after a pre-set time, to be acknowledged again. The acceptance should be part of either an alarm or event log, whichever is applicable. The graphic colours during normal plant operation should be acceptable to the eyes for the long periods that the operators have to use them, since normal operation means attention should not have to be drawn to any parameter. Colours can be used to differentiate between different types of operation, service, product, state, etc. It is worth considering different background colours for different areas of plant, which will give operators a quick reference to the display from a distance. As you can no doubt imagine the combinations are numerous, but ensure the same standards apply to the whole plant. This will ensure that with appropriate training any operator can be comfortable with operating any area of plant with a minimum risk of mistakes.

8.10   Mimic diagrams

On some applications mimic diagrams are still favourite for interfacing PLCs to the operator for control and information. The great disadvantage of mimic diagrams is their lack of flexibility to easily accommodate change of process design, etc. even with 'tile-style' mimics. Where the design is fixed, however, they do have a place. When using mimic diagrams the inputs from push buttons and switches, along with the outputs to indicator lamps, and devices, have to be driven, normally by the PLC. This requirement must be taken into account when sizing the PLC system since it will generally be required to interface the mimic and drive lamps, etc. as well as accept push-button inputs, etc. A good interactive mimic diagram can easily double the I/O count of a PLC system, which obviously has significant project cost implications. Nowadays the mimic diagram is generally being superseded by the video terminal and keyboard. Using a mimic diagram to operate a plant which uses a DCS for the plant control will use a large amount of contact (digital) I/O on the DCS, and will hence be very uneconomic. If it is necessary to interface a DCS using a mimic diagram, the most economic solution is to use a PLC, dedicated to the interface of the mimic diagram. The PLC is then directly linked, via a communications port, to the DCS. Bear in mind the earlier comments about update and action speeds if this combination is to be employed.

8.11   Specification

No matter what type of control system is to be used, as has been emphasised elsewhere in this book, the need to accurately specify the controls required for a project is even more important when a supplier is providing part of or a total package involving control and display equipment. At the stage of generating the enquiry for suppliers to tender against, all aspects of the project control requirements must be fully detailed. These details should include the nature of the measurement equipment, final control elements, type of process and type of control loops. For any control system supplier a fully detailed process flow sheet/P&I diagram is essential. The process flow sheet/P&I diagram is the ideal start point for interactive graphics displays, and the control system supplier can fairly accurately estimate the graphics content of operator displays, etc. from this. Additionally the supplier requires detail of how motors are to be started and stopped, e.g. singly or in groups, and how the control system is to interface to motor control equipment as well as management information systems. All aspects of any equipment to be interfaced to the control system, and how it is to be portrayed in the displays, must be specified, e.g. the process operating philosophy as detailed above. In many cases there are standard packages available from the system supplier, but these must be identified and confirmed as satisfactory for the project requirements. Such things as alarms may not need to be fully specified at this stage, but any information as to what is required overall should be included in the specification. Most important of all is the environment in which the control equipment is expected to work. It is no earthly use expecting control equipment needing an air-conditioned room to work if it is to be located on a production floor exposed to wet, vibration, dust and wide variations of temperature. Specify where each piece of control equipment is to be located and the type of environment it will experience. It may not just be a question of the equipment operating properly in the various locations;

safety may also be a factor. Any electrical or electronic equipment not designed to work in wet environments is potentially lethal to a user. All aspects of what is required of the control equipment, and where it is to operate, must be specified if such dangers are to be avoided. If you are not sure of the final location of electrical or electronic equipment, always assume that the environment will be the worst to be encountered on site! Although this may incur additional costs, remember *safety is of paramount importance*. The supplier will generally assume the equipment will be located in a good working environment, which can be satisfied by lower cost cabinetry, etc. not suited to humid, wet, hot or dusty locations. On many process control applications operator terminals, etc. will be subject to humidity, dirt, dust and general grime, not suited to low cost, low specification cabinetry. Remember the golden rule: *you get what you pay for*, and *if it is not specified, it will not be included*.

8.12   Trending and alarms

Trending and alarm facilities are generally available on all SCADA and DCS packages, but recorders will be necessary where mimic diagrams, single- or multi-loop controllers are employed. Alarms will also require a separate unit when mimic diagrams are used, and in some cases a separate specialised alarm package may be required for dangerous processes, even when SCADA or DCS systems are used for the control of the plant. Today there are a range of chartless recorders available which may be worth considering for the trend recording requirement. These units save the problems of chart changing, pens running out, etc., and most have a memory facility which can be down-loaded onto a PC for analysis if required. Most modern controllers are able to generate alarms, sometimes needing extra cards to be fitted to give the facility. These alarms can then be hardwired to an alarm display unit. In the case of recorders many also have alarm facilities which will give a contact output to generate an audible alarm via a klaxon, etc., but will also mark the recording where the alarm occurred, which can be very useful when diagnosing process problems later.

8.13   Non-disclosure agreements

Many projects involve process details which are very confidential. Whenever this occurs process/project security should not influence the way equipment is supplied. Sometimes it is felt that to spread the supply of equipment over several sources will maintain security of confidential information. Unfortunately this is more often than not the easiest way to make information insecure, and leads to operational problems of the controls due to essential information not being given to a key supplier. Most suppliers are used to complying with non-disclosure agreements, and the fewer the suppliers involved in a project the lower the chances of leaks of information. If project/process security is of utmost importance then a non-disclosure agreement should be signed up prior to any specification being passed to any potential supplier. This will also act as a filter of potential suppliers, since any who are not prepared to sign such an agreement should be dropped from the potential suppliers list at this early stage. A non-disclosure agreement can work both ways for the supplier and the customer. Development of specialised control equipment is essential for any supplier to survive, and to be able to work closely with a client in the development of control strategy, and its application to the client's process, while being protected by a non-disclosure agreement, can benefit both parties. Most leading control equipment suppliers and manufacturers develop much of their control equipment in this way. I have been involved in several such scenarios and can vouch for the benefits gained by both supplier and client from such joint non-disclosure agreements, and development of the equipment. The supplier's understanding of the client's plant, process problems, etc. and the client's understanding and appreciation of what the control system can do, and how it does it, are way beyond the levels experienced in a normal supplier/client relationship where no such agreements exist.

9 Compressed air

In Chapter 3 on final control elements, and control devices, we mentioned the use of compressed air as one of the most effective means of powering actuators. In large factories and process plant the compressed air supply system is one of the essential services required for the operation of the plant. Even with the ever increasing use of computers, the power to control the actuators is still generally compressed air. The air supply may not be your responsibility, but if the correct quality, at the correct pressure and volume, is not available when the control device requires it, the result is control failure! It is in the interest of the control system specification/application engineer to investigate to ensure that the air supply is adequate to meet all the needs of the controls.

Compressed air is very often taken for granted and the result can potentially lead to control system failure. In previous chapters there has been reference to good quality compressed air for supply to pneumatically operated control valves, current to pneumatic signal converters, solenoids, etc. On most of the applications of compressed air for control purposes the requirement is for clean, dry air and where cylinders are used, lubricated air may sometimes be needed. The most common air supply pressure encountered on plant is in the order of 7 bar (100 psig). For most signal converters, control valves and pneumatic transmitters, however, this supply has to be reduced to suit the equipment. Normally the requirement is for a reduction to 2 bar (30 psig) approximately. The use of air filter-regulator units will generally satisfy the requirements, and it should be sized according to the maximum needs of the pneumatic equipment to be supplied. If equipment is provided with an air supply pressure exceeding the specified maximum, damage to the equipment will invariably occur, and in many cases the damage will be catastrophic!

9.1   Air compression

There are three common types of air compressor employed for providing plant compressed air supplies: reciprocating, screw and rotary vane/blower. The small and medium plant units are generally reciprocating, with screw types becoming more common for medium and large plant requirements, whereas the blower, or turbo, types are generally to be found where large volumes of compressed air are needed. The most common drive used is electric, but steam turbine and diesel engines are often employed. For very small supply requirements single-stage compression is generally employed, but for others multi-stage compressors are used, two-stage compressors being the most common. Single-stage is the least energy efficient, with two- and three-stage giving improved energy savings on compression, but requiring more complex plant, with the added costs of manufacture and maintenance. The action of compressing air from atmospheric pressure to 7 bar generates considerable heat no matter which type of compressor is used. Efficient cooling of the compressor is essential, and in many compressed air packaged units this heat/cooling is used to dry the compressed air. Compressing air not only generates heat, but all the moisture in the inlet air is compressed along with the air. This moisture will drop out as water when the compressed air is cooled down, since the higher the pressure of the air the less water-carrying capacity it will have. Generally the air leaving the compressor will just be able to carry the water it contains, but any drop in temperature will result in the water condensing and 'dropping out' in the pipework or the equipment. This water is very detrimental to the pneumatic equipment being supplied with the air and should always be removed. It will also cause corrosion on the inside of galvanised pipework where threads are cut and hence exposed to the entrained moisture in the compressed air. On many occasions air supply pipework which looks perfectly serviceable externally suddenly breaks where the water has rusted the pipe from inside. Not only is this inconvenient, it can be highly dangerous. To minimise the problems caused by wet air it is desirable to remove the entrained moisture as soon as possible after compression of the air. There are several methods employed to dry the compressed air, one being the use of the heat of compression to assist drying by passing it through a 'heat wheel'. This method is employed on some medium and large compressor plants, but is generally not found on small units. Other common methods of drying compressed air include refrigerant dryers, and desiccant dryers. Refrigerant drying lowers the compressed air temperature well below the lowest dew-point expected, and the resulting water dropped out is removed by drain traps. Typically a dew-point of 30°C below ambient is aimed at, as this will ensure that the compressed air supply will be sufficiently free of moisture for most applications. Desiccant dryers are usually of two chamber construction, each filled with inert water absorbing desiccants. The air is passed through one chamber while the other is heated, and vented, to dry it. After a pre-set time the chambers are changed over and the air passed through the dried desiccant while the other chamber is heated, vented and hence dried.

Air compressors should be sized to ensure that the compressor is loaded for approximately 50% of the time, with a stand-by unit available for automatic start-up in the event of failure of the first, or an excessive demand. When sizing compressors the maximum and normal demands should be specified to the supplier, along with any expected peak demands, e.g. plant start-up. It can be very embarrassing if the compressor is sized for the normal plant demand but cannot generate enough compressed air to actually get the air main charged if the supplied equipment is left turned on! You might scoff, but it has happened, often!

9.2   Oil-free air

Like water, entrained oil in compressed air is also detrimental to most pneumatic equipment. Removal of water is normally regarded to be of prime importance, but so is the removal of any oil. Some compressors are 'oil-free' by design, but many are not. If there is any possibility of oil being present in the compressed air it should be removed before any air drying equipment is encountered by the air on its way to the air receiver and then the plant. This can be done by oil-separators installed immediately after the compressors, along with special filters installed following the separator. Oil separators remove the large droplets of oil, and the filters will filter out the smaller particles. The obvious disadvantage of air compressors which are not oil-free is the higher maintenance required for the upkeep of the filters to ensure an adequate quality of the compressed air supplied to the plant. Oil separators, on the other hand, are normally fitted with timed blow-down valves (5 to 10 seconds every 10 to 15 minutes); they generally do not need a large amount of maintenance apart from ensuring that the vents are clear and the valves are operational. Timed blow-down valves should also be installed on air receivers, in place of the normally accepted automatic 'float' type water traps, to ensure that any oil which is left in the air and which may 'drop-out' in the receiver is blown out. The reason for this change is that the small ports in automatic water drain traps will be blocked by entrained oil, especially when emulsified with water.

9.3   Air receivers

To overcome the surges experienced on plant and to avoid overloading the compressors or causing large falls or fluctuations of pressure, an air receiver should be installed. Normally such a receiver should be sized in proportion to the installed air compressor capacities and the plant demand. All air receivers should be fitted with automatic water drain traps and these should always be installed after the dryer, etc., on the final outlet to the plant. Additional receivers can also be installed at the points of high surge air demand, to ensure that the air pressure throughout the plant is maintained at sufficient working pressure at all times. Sizing of air supply mains is critical if air pressure loss is to be minimised. It is always worth installing pipework one or two sizes above those calculated, to accommodate future plant expansion, and to ensure adequate pressure at the extremities at all times. Strategically located air receivers will improve air supply pressures on a plant where capacity has grown above the air piping design capacity, but these should be regarded only as interim measures, prior to new increased sized piping being installed. A ring-main is also a worthwhile investment since this can ensure that much supply pipework maintenance, etc. can be done without shutting down all the air supply system and, in conjunction with strategically placed air receivers, can keep the plant running in many situations of air pipework and equipment maintenance, etc. Another consideration is to provide any air receiver with 'lockable' by-pass valving and pipework, to allow for insurance inspection during plant running periods. Maintenance and inspection on a well designed and thought out air supply system should be virtually transparent to the running of plant. Remember air receivers are pressure vessels and are hence subject to legislation regarding the fitting of safety valves to suit the receiver capacity and certified gauges, and they must have an annual inspection by the site insurance inspector.

As a guide to air receiver sizing one compressor manufacturer gives the following method of calculation:

Receiver capacity m^3 = (Free air requirement m^3)/(Allowable pressure drop in bar)

The allowable pressure drop is the difference between the normal air supply pressure and the minimum air pressure required to operate the pneumatic equipment. For receivers fitted to air distribution systems a guide pressure drop for a 7 bar air main is 1 bar at full air supply volume (free air requirement). Generally air receivers are available in a series of standard sizes. Calculate the size required and then choose the next size up. To satisfy large requirements a combination of receivers is quite acceptable, but each must be fitted with a safety valve, pressure gauge, drain equipment and, if possible, individual isolation/by-pass valve arrangements to allow for easy inspection during plant operation.

9.4   Air distribution

It has already been pointed out that compressed air contains a high level of moisture. Even if this is removed by various types of dryer, there will possibly be a little remaining which will drop out when the air supply main passes from hot to cold areas. Air mains should therefore be installed with a 'fall' to a trap/drain point, to allow any water to be removed. All connections to the air supply main, for providing any equipment with air, should be from the *top* of the main. This may not totally stop water being entrained in the air supply to the device, but will minimise the chances of water being passed to it. Employing a filter before the pneumatic equipment will further reduce the possibility of water entering the equipment. Where the equipment employs small orifices the filter needs to be fine to stop any dirt or water blocking the device, and causing it to fail. Typically standard air filters are 25 micron. These will be sufficient for general purpose use, but on such items as I to P (current to pnuematic) converters, fluidics, etc. a 5 micron filter is worth considering. To maintain availability of plant, filter-regulators should never be run at their maximum capacity. Sizing filter-regulators for a maximum of 50% loading will give high dividends in pneumatic equipment reliability, to a much greater degree than with any other service. In most countries all air supply/distribution pipework is subject to legislation regarding the standards of equipment used including pipe, fittings, isolation valves, etc. Where this legislation is in place the complete air supply/distribution system must also be fully documented.

The way the air distribution round a plant is installed requires some basic techniques to be employed to get the most reliable service from pneumatic equipment. We have discussed the need for air receivers and the advantages of a ring-main, but there are other considerations. The size of the air main pipework is of utmost importance. If the bore is too small a high pressure drop will result when air is being used. This pressure drop will result in a low delivery pressure at the point of use, and will waste energy. The generally accepted method of sizing air distribution mains is by the maximum air velocity expected when the air consumption is at its maximum. The maximum velocity should be no greater than 9 metres per second (9 m/s). For practical application on new plant it is better to consider 5 to 6 m/s. This will allow for considerable extra demand which invariably will apply as additional equipment is installed during the life of the plant. At air velocities above 10 m/s any entrained water will not precipitate out, and hence will pass directly to the point of use. As detailed above, every take-off point from the air main should be vertical and fitted with an isolation valve as close as possible to the air main. This will ensure that any small amount of water present in the air will generally stay in the main itself, hence avoiding its passage into delicate pneumatic equipment. The isolation valve will ensure that any problem with the pipework from the take-off can be isolated. This will enable repairs to be undertaken, and will avoid shutting down the air distribution main which could cause disruption of plant operation. Additionally every piece of pneumatic equipment should be fitted with its own isolation valve. This may sound very logical, but many plants do not have this standard and suffer many hours of disruption to production when equipment fails or modifications are required. Also if local isolation is not fitted it is impossible to carry out running plant maintenance on pneumatic equipment which may not be in operation all the time. Finally, as emphasised earlier, all air distribution and supply pipework must be installed so there is a small gradient, or fall, built into the installation. At the bottom point of each fall there should be fitted an automatic drain, with an isolation valve before it. This will ensure that any water not removed by the dryers will be removed if it precipitates out.

9.5   Compressed air cost

A ready supply of compressed air is, more often than not, taken for granted. No matter which type of compressor is used the power required to compress air is high. Theoretically for 100 litres per second of air, at a pressure of 7 bar, at least 24 kW of power, on two-stage compression, will be required to provide the air, plus extra power to dry it! On a typical medium sized compressor plant I have found 117 to 120 W per cubic foot per minute (cfm) of clean, dry air at 7 bar to be the power needed. On this basis a typical pneumatic transmitter, level or pressure, will be equivalent to a 60 W bulb to run, i.e. 525 kWh per year, and a 100 mm line size control valve can be as high as 400 W if it is modulating continuously, and 100 W for steady state. When selecting pneumatic equipment it is worth considering the air consumption as one of the deciding factors in the choice of apparatus, as this can significantly affect the running costs of the process controls.

9.6   Dirty air

The need for clean dry air supplies to pneumatic control equipment cannot be emphasised enough. Although the effects of dirty, wet and oily air may not be immediately apparent when a new plant is started up, they can be very costly long term. Process plant may suffer poor control initially, but if the dirty air supply is not corrected the plant may have to be shut down to allow for all the pneumatic equipment to be cleaned out and air mains, etc. to be cleaned. Fitting air cleaning and drying equipment at this stage will not only be costly, but there will still be a residue of dirt remaining which will take time to disperse, and may still cause problems before it is finally cleared. I have experienced just this situation and, after cleaning all the pneumatic equipment and fitting oil separators, filters and dryers on a plant, supplied by one duty plus one stand-by 150 cfm compressors, it took over two years before the oil was finally no longer causing major problems. Remember: *dirty air means poor control, low reliability, and increased maintenance.*

9.7 Air cylinders

For many applications **air cylinders** may be employed for on-off control, e.g. spade valves, or where other types of movement are required, e.g. diverting products on a conveyor, lifting, pressing, etc. In these cases different types of air cylinders are often used, ranging from conventional cylinders with piston rods to rodless cylinders, pancake or flat cylinders and sometimes air-rides or bellows. They are one of the most cost-effective means of obtaining the necessary force and action and give easy automation when controlled by solenoid valves, etc., linked to the control system, or supplied with piloted regulated air, fed from a signal converter such as an I to P where a varying force is necessary. Modern air cylinders, unlike the units of 20 years ago, do not often require lubricated air, since modern materials, such as PTFE, are used for sealing rings, etc. Although lubrication may not be necessary, clean, dry air is essential, and to a much higher degree than for their old counterparts. The working life of modern air cylinders will be reduced significantly if the supply air is not of the best quality, with consequent premature failure. In the case of air-rides or bellows the rubber construction makes the supply of oil-free air essential.

9.8 Solenoid valves

Solenoid valves are one of the devices used to convert an electrical switched output from a controller, such as a PLC, to pneumatic action of a cylinder, or similar device. They can also be used to turn supplies such as air, water or gas on and off. Solenoid valves have a small number of moving parts in them, and hence require the correct air supply when used on compressed air services. All solenoid valves used on air must be supplied with clean dry air. On some applications where lubrication is required, e.g. air motors, an in-line lubricator is needed. To ensure correct, reliable operation of associated solenoid valves check if the solenoid valve is suitable for use on lubricated air. Some solenoid valves have internals which will be degraded by oils. Solenoid valves are particularly vunerable to dirt in the compressed air supply. Small dirt particles are abrasive and will cause wear and tear to the fine 'O' rings fitted inside causing premature failure of the device.

9.9 Summary

This chapter may create the impression that pneumatic control equipment should be avoided due to the need for the necessary degree of cleaning of the air, and the cost of supplying it. In many applications, however, pneumatics are still the most cost-effective, reliable and cheapest to run, and in some cases the only effective, means of obtaining the necessary actions.

The reason that this chapter has been included is to highlight the need to assess all the control equipment required, and to ensure that there is a full awareness of potential problems. Addressing these potential problem areas at the design/installation stage of a project can ensure a very successful process control system, in both the short and long term. On some installations I have experienced old pneumatic transmitters connected to pneumatic to current (P to I) converters and then the signal connected to an electronic controller input. Although adequate for the purpose, for applications in some hazardous areas this may be the most desirable method. However, for general duties, as can be seen from above, the most effective solution is to employ a modern electronic transmitter. Not only will this reduce running costs, it will also improve measurement accuracy and reliability (one less device to go wrong!), along with reducing maintenance. Today's electronic transmitters have typical accuracy of ±0.25% or better, with many giving ±0.1%, and ±0.05%, whereas pneumatic transmitters have accuracy in the order of ±0.5%, and many are no better than ±1%. Even when pneumatic transmitters are provided with clean, dry air the unit must be regularly serviced, and calibration should be checked at least once per year, which requires removal to a workshop. Modern electronic transmitters need a calibration check once every two years, little or no service and, with SMART™ versions, removal from plant is not always necessary.

9.10   Hydraulics

There is no separate chapter devoted to this subject, but since many of the actions available with hydraulic devices are similar to pneumatic devices, it was felt worthwhile mentioning it here. **Hydraulics** utilises devices such as cylinders, solenoid valves, etc. which give similar action to the pneumatic equivalent devices. Hydraulics, however, gives a much higher level of available power for the size of device. Unlike pneumatic devices hydraulic devices are normally sited with their own hydraulic power-pack nearby, or are even integral with it. The hydraulic power-pack is normally a small hydraulic oil tank fitted with one or two hydraulic pumps, filters, etc. Like pneumatic devices, those used on hydraulic systems are very susceptible to dirt and water. It is essential that wherever possible the hydraulic power pack is sited in a clean, dry area, and that all fillers and vents are kept clean. Hydraulic engineering is quite a broad subject in itself, but for purposes of this book it can be likened to pneumatics with respect to the way the control is interfaced to the control system, either PLC, DCS or other.

10 Choice of system and installation

The previous chapters have identified most of the factors governing the choice of equipment for the different control requirements of the process to be controlled. We have also identified such requirements as management information, operator facilities, etc. so we should be able to select the measurement devices, and the control devices, having also identified and taken into account whether the process requires cyclic or continuous control. The loops under consideration are those for on-going control of the process, not necessarily the motor control, unless stopping and starting of such items is part of the cyclic or continuous control strategy. For small control requirements this may be satisfied by single-loop controllers, etc. but for a large control requirement, or extension of an existing one, the selection of equipment takes on a different emphasis. Where there is to be an extension of existing control facilities the choice is limited to existing site standards, or equipment which is compatible with existing standards employed on the site. With the advent of the new communication standards the level of choice in this arena will become almost as great as a new control system type selection project, but this will be a few years coming. Where equipment manufacturers are working to the same communications standard the compatibility must still be verified by the prospective supplier since, like any software, the release levels may differ and hence compatibility may not be exact. This check will also apply to extensions of existing control equipment as once again software levels may not be fully compatible. Often to go to the current levels of software to enable the most up-to-date equipment to be used may require phasing out some of the older, existing hardware. It is therefore essential to detail all existing equipment to be interfaced in the system specification for any extension, since the functionality of the control system as a whole is of prime importance, and the detailed specification of what the new equipment is expected to function with should be made the responsibility of the supplier if at all possible. It is therefore of vital importance that the records of the original equipment supplied, and all changes made to the control equipment, are kept up to date. If records are suspected of not being accurate a survey of the existing equipment can be carried out by the prospective supplier, but often this will incur an on-cost to the project.

10.1　Totally new system

If the choice of equipment for a sizeable control system is totally open it will generally be determined by the type of control required for the process. If the majority of control loops are of a cyclic nature the choice will most likely be for a Programmable Logic Controller (PLC), or networked PLCs. In Chapter 5 we covered PLC control in some detail, and outlined the information which it was advisable to include in the equipment specification, and the use of a Systems House for advice, design, etc. With PLCs the plant geography will determine the number of I/O necessary in each area to satisfy the process control needs and good plant segregation. A careful study and consultation with the process design team will identify the sections of plant which can be run independent, or partly independent, of each other. To improve segregation may require the incorporation of enlarged holding tanks, etc. to achieve a 'process break', but it may be worthwhile in order to give good overall plant control, plus improve process and control trouble shooting. A continuous process can quite often have incorporated holding tanks, etc. which can be used to give a few minutes of plant section down time to carry out emergency changes, repairs, etc. and these points will be the natural 'break points' between one PLC and the next on a networked system. Each PLC can then be programmed to run in isolation if necessary, which will make a fail safe programme easy to achieve. Additionally if communication over the network is lost the sections can be programmed to shut down automatically in a controlled and safe manner. Study of many existing process plant control systems will show this philosophy is well used in all types of control systems including PLC, DCS, and plants using single-loop controllers. The philosophy outlined will also identify where operator access to control is required, and will pre-define the network architecture. As already mentioned, an ideal location for the PLC I/O associated with motor control will be in the motor control panel, or switch room. This may also be an ideal location to serve the field equipment since similar cable routes can be used, provided segregation of signal and power cables is carried out (see installation specification).

Where the majority of control loops are continuous the choice of control equipment will most likely be for a Distributed Control System (DCS). Motor control can be easily accomplished on most DCS, and therefore should not be considered as part of the make-up of the control loops required unless a large amount of cyclic running of these items is part of the control philosophy, as mentioned earlier. The architecture of any DCS is pre-defined, but is similar to the architecture of networked PLCs, as can be seen by comparing Figs 5.5 and 7.1 in previous chapters. The control equipment cabinets are again located in a similar manner to the PLC I/O, etc. and these locations will generally be defined by the plant geography. Again observation of process plant sections will assist in fail safe plant shut-down engineering, etc. The location of cabinets can again follow motor control location, but more attention to possible interference from electrical power may be required. In general DCS can provide all essential control features for both continuous and cyclic plus motor control, etc., and in addition can give recording, totalisation/integration, reporting and other management and operator information as part of the basic system. The pre-defined DCS functions will invariably contain facilities to integrate motor/pump control with continuous control, without need to revert to any other control system. With PLC systems the additions identified above will mostly be additions to the basic PLC, as will operator interfaces, etc. This may require equipment from different manufacturers to be interfaced together to achieve the required result with consequent problems of compatibility, maintenance, etc. With different DCS manufacturers the field equipment cabinet/I/O capacities may vary and this should be one of the factors considered when making a choice of supplier. With the majority of suppliers, however, the I/O make-up within the field equipment cabinets is very flexible, and can be adjusted to suit most control requirements. The most influential factor in the decision process for the choice of DCS is if configuration/re-programming is required on running plant, and to what level re-configuration will be required. All systems will allow 'tuning' of control loops to be carried out, but some will not allow any other functions to be accessed on the system when running. The systems requiring shut-down to configure/programme, etc. may require the process to be off-line for several hours to accomplish the desired changes, which may not be acceptable. Investigating and considering these factors, if they are important, will considerably reduce the number of manufacturers and systems under consideration. *With any system where changes to configuration on running plant can be carried out it is essential that the 'configuration mode' is only accessible to qualified, trained and authorised personnel.*

10.2   Interfacing 'new' with 'old'

Where a new project is required to be controlled from existing control terminals, or interface to existing plant controls, the compatibility of control equipment must be considered, even if the new equipment is of the same manufacturer! With technology advancing so rapidly a control system can be superseded even before it is finally installed and commissioned! This does not mean it will not work, or not be supported by the supplier, but that due to more advanced technology being incorporated in the components, the equipment capabilities will advance along with the application of these components. A couple of years between phases of development of a plant can span several advances of the control system, from the existing control system employed to that available at the outset of the next phase of the process plant. Sometimes the hardware may have changed little but the software may have been significantly advanced. When embarking on a support agreement with a supplier it is therefore a great advantage to include software up-grades with the support package. This will overcome the substantial step-change which may be encountered when expanding a system a few years from the original installation. In some cases the 'old' will not readily interface with the 'new', and at worst it will be uneconomic to do so. It may be necessary to use an extra piece of hardware to interface the 'old' with the 'new', along with some specially written software, to complete the communications of the interfacing. It will often require some of the 'old' control equipment to be up-graded, and this will have to be accommodated in the costing for the project. Precise specification of such interfacing is vital and it may be worthwhile making the supplier of the 'new' equipment totally responsible for this interfacing as part of the scope of supply. To specify that the interfacing with existing equipment will be 'transparent' may not be possible, and could result in a very high priced quotation. If a problem of interfacing is identified, and the practical requirements are carefully thought out and discussed with plant management, operators, maintenance personnel, etc. a cost-effective compromise can often be achieved.

10.3   Compatibility of control equipment – systems integrator

Whenever more than one source of equipment is to be integrated on a plant the problem of interfacing arises. Unfortunately very few manufacturers provide equipment that will readily interface with another manufacturer's products. At the time of selling their equipment a supplier will generally say that their equipment uses standard communications and interfacing should be easy, but more often than not, the term 'easy' has been somewhat exaggerated! The communications standards may well be the same, but there are many factors which may differ, necessitating special interface drivers to be written; special hardware may even be required. At best there will be an on-cost, and at worst the equipment may never interface successfully to satisfy the client's specification fully. It is always worth requesting from potential suppliers a list of satisfied clients, where similar interfacing to your project specification has been carried out successfully. Never be afraid to contact those clients for their comments. Their remarks may save you a major problem later in your project, and steer you away from potential disaster. On large projects it may be worth considering employing a reputable Systems Integrator (SI). A Systems Integrator is a highly skilled, experienced person, who is capable of ensuring the compatibility of equipment and that all items will integrate. For obvious reasons it is essential to employ such a person early in a project so that the minimum amount of effort, hardware, software and hence cost is incurred to make equipment work together. GAMBICA, the association for the instrumentation, automation, and control industry, has produced a multi-point plan to assist in the selection of such an SI. In choosing a Systems Integrator it is essential you ensure that he or she is fully conversant with the equipment to be used and, if possible, the type of process. This will be very hard to achieve fully, and such a service will obviously not be cheap! Therefore such a person should only be used for a very large project. However, if you take the time to specify the control requirements of the project in detail and, if possible, single source the control system, the need for such a service should be minimised, if not completely avoided.

10.4   Update and action speed

One important factor to contemplate when considering overall control performance is the rate at which process changes are updated on the displays, and how long any action made through a terminal takes to be carried out on the final control device. If several interfaces are involved, the scan speed of each becomes cumulative on the overall update and action speeds. What seems to be a few milliseconds at each stage can result in seconds when parity and error checking, etc. are included. With improvements in communication speeds this is becoming less of a problem, but it can still be annoyingly, or at worst, dangerously slow in some instances. When drawing up the operation specification of the equipment ensure that the maximum allowable time of update, and control action is identified and clearly defined. This is of prime importance on fast processes where delays can cause dangerous situations to develop. Boiler control on 'water tube' boilers is one such process where any delay in pressure or level control can have dangerous consequences. It is true that boilers have safety cut-out devices, but the control should never cause these to have to come into effect. Remember safety devices such as safety valves and low water trips are there for ultimate protection only. Once a safety has been lifted a couple of times it will invariably pass a little steam when shut, which is a waste of energy, and will ultimately lead to 'wire drawing' across the valve seat so increasing the rate of leakage. A badly 'wire drawn' valve seat will normally require premature, costly and inconvenient replacement of the unit, for which the control system designer will get no thanks.

10.5   System specification

For any application a detailed specification of the scope of the control requirements, and operator terminals, etc. is essential, as we have already identified in previous chapters. The specification should also include the plant geography, the operating environment and full details of all other equipment to which the control system will be interfaced. The nature of what is expected of the supplier should also be detailed, e.g. equipment supply only, turn-key project, etc., and a summary of what you, the customer, expect to achieve at the completion of the project. The majority of suppliers should be able to give a detailed quotation of their equipment to satisfy the requirements from a plain English specification; in some cases site visits and discussions about the project will be necessary for an accurate appraisal of what is needed. The supplier should give details of how the equipment will meet your specification, plus any additional facilities that the equipment proposed will provide, and a programme of stages of construction, inspection and delivery to site. The training necessary should also be identified by the supplier at the quotation stage, since this will no doubt cause some disruption to production if existing personnel are involved, and this will need to be included in the project plan. If training is the responsibility of another department, they should be included at this stage. If a supplier does not respond in the fashion detailed above seek the reason for this, and ask for a more detailed response to your enquiry if you are not familiar with the proposed equipment. If this is not forthcoming the most likely reason is that your specification is inadequate, or has been misinterpreted by the supplier. Other reasons may be the supplier's lack of knowledge of other equipment to which interface is required, embarrassment that the current models will not readily interface to what is on site, or not being familiar with your process or methods. If the enquiry is responded to by several suppliers the response should be similar from each if the specification is adequately detailed; consequently the proposals submitted can form a good method of primary selection of supplier. Another factor included in the specification should be your requirements for supplier support, and expected response to site, of each level of service, when problems arise, including the time of day, etc. when support may be called for and how this will affect the response. The response times and costs may also influence the selection process, e.g. if the process is 24 hours per day, 7 days per week, a call for breakdown assistance on Friday cannot wait till Monday for attention, nor can you afford the cost of an engineer coming from the 'other side of the world' every time you require any level of support. Full details of supplier support contracts, costs, etc. should accompany any quotation for major projects, and if possible should be included for the first year or two of operation of the system. It may be worth specifying when such support is expected to commence as most suppliers will deem the start to be 'delivery to site' not time of commissioning/acceptance, unless this is noted in the contract. *Again if it is not specified it will not be included!*

10.6 Assessing a system to suit the application

In choosing any control system there are many factors to be considered, and the decision can involve considerable emotion which could cloud the issue. Before embarking on the choice of supplier every attempt should be made to remove unnecessary influences which could result in the choice being wrongly biased. For every application a specification of requirements can be developed, but this alone will not fully identify the right control system for it. One method of reducing the number of systems to consider is to draw up a table for comparing the systems under consideration. A series of factors important for the application should then be added to the first column of the table, with each factor having a total score in proportion to its importance, arranged so that the total score of all factor maximums added together will give 100 say. For each system under consideration include a column, and *only* from *published data* supplied by the manufacturer compared to each factor in the first column scored for each. If all the control system totals at the bottom of the table are compared a comparison score should result, the best choice for the application in question having the highest score! This is one method of comparison which should be free of sales gimmicks, promises, etc. and give an evaluation which can include all personnel and their requirements in an easily understood format. There are many other methods to assess the most suitable control system, but if an unbiased assessment can be made and this results in the most favoured system also identified by other assessment types, then the right control system for the job has been well researched to prove the choice. Such a method of comparison will no doubt raise questions about all the systems under consideration which will be of benefit to all those concerned with choosing the system, since a better understanding of what each supplier is offering will result. This method of selection will also do a large amount to make you familiar with what each system can do, and how it does it, along with the capability of each supplier. It will also identify the suppliers who will give assistance at this early stage, and hopefully the same level of help will be repeated throughout their involvement if they are chosen. With a large project the sales team of a prospective supplier will normally need to involve some of the company project team, giving you a chance to assess them.

10.7 Future requirements

When selecting a control system it is worth considering what other, non-control, future plant requirements are, and if possible developing a systems strategy encompassing all information technology/management systems requirements, etc. This may seem a tall order but consideration of such plans at control system equipment selection time could save the embarrassment of the selected controls not being compatible with future installations of information systems, etc. With the availability of the new communications standards any choice of a large control system should take this into account, and systems not compatible to any of these communications standards should not be considered, unless special circumstances prevail. Many system suppliers will normally identify their product capabilities in these areas before you ask the questions, but any future requirements you can identify are additional features to aid the selection process.

10.8   Terms of supply and delivery of equipment

Do not leave the 'terms of supply and delivery' to others. Ensure that *you* are satisfied with all the terms and conditions of the contract before it is signed. There may be factors which will be your responsibility and affect the supplier's commitments. In many cases the off-loading of the equipment must be done in a specified manner, or supervised by a representative of the supplier to satisfy the terms of the contract. The initial powering up of the equipment must be done by the supplier in many cases, and the tests to be carried out for acceptance must be done at a specified stage of factory assembly or installation. In many cases the standard terms and conditions of supply will not totally suit the project, and any modification to these must be fully agreed by all parties before an order is placed. This may be another factor in the choice of supplier/equipment, since delivery times, etc. may be very critical in the project programme. If there are any details needed early for the installation of the equipment this must be made known by the supplier at the time of order placement, not when the equipment is delivered. Such information availability must be established in the enquiry specification to ensure that it does not create a problem in the project plan. In some cases a 'factory acceptance' has to be completed before delivery will be made. This may incur overseas travel not budgeted for, or may need somebody competent to deputise for you if you are not available. It is also essential that you identify that factory acceptance does not reduce the quality you are expecting of the equipment guarantee. It is highly possible that simulated conditions at the time of factory acceptance do not truly represent what will be experienced on site, etc., or that the software and equipment work very well on their own but do not perform to specification when linked to the plant, or other suppliers' equipment and software. All the points raised here may seem simple but they can cause major problems if not identified at the time of placement of order.

10.9   Enquiry information

To summarise, and remind ourselves of the details required to be included in the enquiry specification for a control system of any type, it may be worthwhile fully detailing all the information under the following categories.

1. *Control requirements* Number of devices under control, type, etc.
2. *Geography of plant* What controls required in which areas, etc., area environment specification (hazardous, etc.)
3. *Operator terminals* Where operator terminals are needed and how many screens, etc.
4. *Other equipment to be interfaced* Any management computer systems, other control systems, etc. and what is the level of interface required (information transfer or control, etc.).
5. *Other software to be interfaced* A small change in one supplier's software can render interfacing impossible. Ensure that all software and the revisions/levels are detailed.
6. *Scope of supply* Turn-key or supply only, etc. (who does programming/configuration?).
7. *Aims and expectations of project* By client! — confirmation to be made by supplier.
8. *Project outline plan* When deliveries, etc. are required.
9. *Training requirements* Operator and maintenance? And to what level?
10. *Installation information requirements* When required, etc.
11. *Guarantees and support contracts* Start dates, scope, response requirements, etc.
12. *Factory acceptance tests* What implications to guarantees, etc. Where? Who attends? When? How will delivery be effected?

10.10 Organising the installation

When the required control system for the application has been identified and ordered, we can start to organise how the system will be installed. The locations of the various cabinets/panels, etc. should already be defined for the various parts of the process plant, and once a system has been selected the cabinet sizes should be known to allow final positions to be established. Also operator control rooms, etc. should already be defined giving the location of operator terminals, etc. Such items as cabling documentation, including routes, connection schedules, etc., will have to be organised. Ancillary services, etc. will be needed, e.g. air supplies, power supplies, etc. The installation contractor will need a *fully detailed specification* of what is required. Accurate and full collation of all the information for installation of any control system is very important to ensure a good installation and an on-going reliable control system. Quite often the 'failure' of a control system is down to poor installation. A good, well detailed installation specification will ensure the success of the project. The following sections in this chapter will give much of the general information which should be included in any installation specification for a control system. Again remember: *you get what you pay for* and *if it is not specified, it will not be included.*

10.11   Communication/data highways

The first task is to define the route for the communications highway. This should be the shortest route, but located away from possible interference from any electrical apparatus, radio transmitters, etc. Typically allow a minimum clearance of 300 mm, plus 300 mm/500 V clearance from all power cables, switchgear, motors, etc. Where the communications highway runs parallel to high voltage power cables (3.3 kV and above) the clearances should be doubled if an alternative route clear of such cables cannot be found. All the communications highway should be run in conduit, or trunking, which is earthed. This should give good mechanical protection of the communications highway. Most communications highways will require special cables, defined by the system supplier, with specific requirements for screen earthing, etc. which must be observed if problems are to be avoided, and guarantees honoured.

10.12   I/O cabling

Having established the route for the communications highway we can now locate the routes for the main cable runs for the system I/O from each of the cabinets. The I/O cables should be kept away from the communications if possible, especially if these are carrying voltages in excess of 100 V (ac or dc). For contact I/O steel wire armoured (SWA), multi-cored cables will normally be suitable, provided the total load is well within the cable specification. For analogue 4 to 20 mA signals, or digital signals, screened cables should always be employed. In the case of 4 to 20 mA multi-cored, overall screened cables can be used with multi-stranded conductors equivalent to 0.5 mm^2 cross section. Many installation contractors, and equipment suppliers may say screening is unnecessary for the 4 to 20 mA signals, but with modern, microprocessor-based equipment, the elimination of all possible sources of interference is essential. The extra cost of running all signals in screened cable will be a good investment. If screening is not employed, and interference is experienced, the cost of retro-fitting all the affected I/O will be many more times the extra cost for using screened cables initially. For digital signals a similar situation exists. For digital signals and low level data transmission twisted pair screened with an overall screen should be used if there is any suspicion of potential interference. As a minimum specification always use at least one level of screening on digital signals. Many computer system suppliers may say only twisted pair cables are adequate. This may be the case in an office environment, but not an industrial one! The same specification for separation from power cables as the communications highway should be observed. It is worth running from the I/O cabinets in multi-cored cables, e.g. 18 or 27 core, to junction boxes located at convenient points in the field, and then running to the 'field' equipment in 2 core, 4 core, etc., to suit the devices. Never run field device power supplies in the same multi-core cable as the signal, unless this is a purpose-made cable, specified and supplied by the equipment manufacturer.

At this point it is worth mentioning cable/core identification. All too often a designer with little or no maintenance experience generates the cable and cable core identification with the result that maintenance becomes a big headache. There are various identification systems in use, but those which do not require a print of the cable schedule for maintenance are definitely the best. This subject can generate much emotion and confusion; however, the best system experienced for DCS or PLCs is one using the 'system' I/O address. All systems require a discrete address for each I/O point. If this address is used for cable/core identification a loop wiring can be traced at any time, without cable schedule documentation, after installation. If this identification is used right up to the field device the correct device can be identified easily by maintenance staff, etc. at any time. Figure10.1 shows an example of using system address for cable and field device identification. No matter what identification system is used all connections of all cables and all cores should be fully identified at every connection point, including junction boxes, etc. to facilitate easy fault finding, modification, etc. at any time during the life of the installation. Whatever system is used for cable core identification always remember that somebody else may have to trace faults, etc., long after the plant has been installed. A core identification system which requires a full set of cable schedules to be able to trace any cores is a nightmare for maintenance staff! Even worse, never employ a system which generates a situation where the two ends of the same cable core have different identifications!

Cable trays, etc. should be well supported and wherever possible should have no more than a 60% loading at the time of initial installation completion. This will allow reasonable capacity for expansion of the control system at a later date, if required. A good installation will last for many years, and a well installed cable network will avoid faults and breakdowns resulting from pulled or strained cables where the cables are unsupported, etc. The support of traywork, trunking, etc. should be increased where there is a possibility of vibration, e.g. on walls adjacent to crane rails, near to heavy machinery, etc. It is worth considering what might happen to cable trays. Where they are large it is highly likely that installation personnel will find it easier to lie on the tray to run and clip the cables than to lean from scaffolding or a ladder. Tray supports also make nice 'grab handles' for operating staff trying to access awkwardly placed equipment, and even good, highly illegal, lifting beams! A few considerations about such misuse should result in traywork being run to avoid these situations, wher-

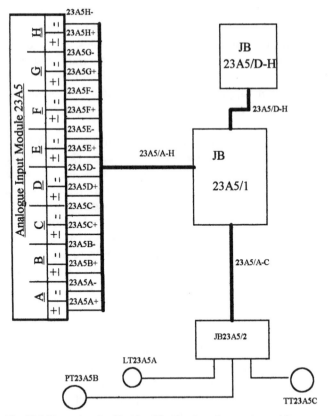

Fig. 10.1 Example of cable identification based on system address.

ever possible. The result will be an installation which will last, and look good, for years.

Where the installation is in an area classified as hazardous the appropriate regulations and guidelines must be observed which may entail some modification to the installation recommendations above, e.g. minimise field junction boxes, etc., but the overall concepts should still apply.

In applications where cables are run high in plant rooms, etc., ensure that the heat encountered at the point of installation is taken into account in the sizing and specification of the cables, especially where they are power carrying. Consider access to such routes after installation. They may be out of the way, but any requirement to run extra cables or repair existing ones will incur the penalty of scaffolding costs, etc. and the delay involved in erecting such means of access. If the environment is dusty, there will be a dust build-up, potential fire hazard and cleaning problem if the tray is horizontal. Positioning the tray so it is in the vertical plane may require more support clips for the individual cables, but will significantly reduce the dust build-up problem, which in many cases would also reduce the problems of fire and cleaning.

10.13   Earthing

Many problems occur from badly or wrongly earthed control systems. The need for a good quality earth cannot be emphasised enough, but unfortunately it is often the poorest part of the installation. Most system suppliers will provide a specification for the earthing requirements for the supplied equipment, which must be observed. Typically any control system will require a 'clean' dedicated earth, independent of any other apparatus, and the system generally should be earthed at only one point in each network area. With most control systems the cabinets can be earthed independently from the system, so enabling all electrical regulations to be satisfied. Typically the conductor size for the control system earth should be twice that employed for the earth for an equivalent sized electric motor, and a minimum of 4 mm^2 conductor size for earth cabling on low power systems. The earth size is not just dependent on the power involved with the electronics, but also the total power which is switched by the various outputs of the system, especially those interfaced to motor control centres, solenoid valves, etc. Most control systems should be earthed in a 'star' formation from the clean earth, but this must be confirmed for the chosen system before installation. Installations in hazardous areas must comply with the appropriate regulations. With signal cables it is always advisable to use screened cables to minimise electromagnetic interference. With all screens the only point of earth should be at the control system cabinet, not at both ends of the cable! Even where low voltages are used for switching and motor control the earthing requirements are no different to higher voltage systems. Initially the earth is for protection of personnel, but also it is there to protect the electronics from interference. The requirement for such a large earth cable is due to the fact that the volt-drop with a poor earth will be high relative to the voltages used on the electronics. Where electronics only require a few millivolts to switch, spurious switching could potentially be induced via the earth leakage volt-drop if the earth resistance is high.

10.14   Ventilation and cooling

In previous chapters we have emphasised the need to observe maximum working temperatures of equipment. If there is a potential problem of high temperatures being encountered circulation fans, forced air cooling or heat exchangers should be considered, according to the cooling requirements. The addition of cooling can degrade the enclosure environment rating if the incorrect unit is installed. If there is any doubt the safest cooling to apply is by 'air-to-air' type heat exchangers mounted on the cabinets. This will maintain the integrity of moisture and dust ingress ratings for the enclosure, provided the units are installed with the correct gaskets, etc. Forced cooling air methods can introduce dust and moisture into the electronics, unless the very finest of filters and air dryers are used, which will increase maintenance requirements. Internal circulating fans will generally not give overall cooling, which may be necessary for protection of the components. Circulation fans are provided to distribute the air within the cabinet, and only prevent 'hot-spots'. Some suppliers fit water cooled heat exchangers to reduce cabinet temperatures. Unfortunately many such heat exchangers are fitted without thought to what happens when they, or the pipework, leak! I have personally experienced the loss of expensive electronic equipment as a direct result of badly positioned water cooled heat exchangers. If they have to be fitted make sure they are at the bottom of the cabinet, without any risk of splash or drips going onto any electrical or electronic equipment. Remember the average water supply pressure is high and a pin hole in the heat exchanger, or pipework can spray several metres! Sometimes the spray may be such a fine mist it is difficult to find, especially if the cabinet internals are not well illuminated. If you can avoid them, do so!

10.15   Control systems power supplies

This subject is most often taken for granted but can be the source of many problems with control systems, remaining undetected for considerable periods and so undermining confidence in a system's capabilities, etc. Most control systems will accept raw ac power and the power supply unit of the system will carry out the necessary 'conditioning', conversion, etc. However, many control systems can be adversely affected by bad quality ac power supplies. It is always worth ensuring that system power is supplied from a secure source that is free from large voltage fluctuations and interference if at all possible. In most cases a back-up supply should be made available, which should automatically be selected on failure of the normal source, giving warning of the status on the control system displays. Where automated shut down of the process is necessary Uninterruptable Power Supplies (UPS) for the control systems should be employed. Such UPS units must be dedicated to the control systems, *including the cooling* of the enclosures. All items of the control system, on the same communications highway, should be fed from the same ac power supply source if possible. If this is not possible ensure that there is no possible link on either live or neutral lines. On many sites neutrals of low voltage (110 V ac) are combined in control equipment panels, etc. This practice can give many problems if carried out on any PLC or DCS installation, especially if the neutral side of the supply is earthed at some point, and should therefore be avoided at all costs. The earthing of the neutral can aggravate the earth problems outlined above. A supply with an earthed neutral should be fed to the system PSU (Power Supply Unit) via an isolating transformer to overcome any possible problems. Another common source of problem is the use of 110 V ac power sources supplied from a 'centre tapped transformer' giving the supply in the form of 55-0-55 V ac. This will give a continuous fault condition on any control system employing 'cabling integrity' monitoring, and will also give rise to earth loop voltages around the system contact I/O, where mains voltages are present. Such power supplies can also cause problems if used to supply PLC programmers, when the PLC itself may be fed from a suitable supply. Ensure the programmer is fed from the same power source as the PLC it is being used to programme, to avoid potential problems. If this cannot be achieved use an isolating transformer for the programmer.

10.16   Measurement and control devices, signal converters, etc. connected to a control system

The measurement and control devices covered in previous chapters are all generally suitable for connection to any control system; however, if non-isolated I/O are used, the measurement and control devices must be themselves fully isolated, otherwise problems with earth loops, non-compatible signal voltages, etc., may be encountered. Any control system will normally be provided with non-isolated I/O, unless specified otherwise! Since many of today's DCS carry out field wiring integrity checks, it is essential that the possibility of earth loops, very high or very low impedance loops, etc. is avoided. In previous chapters the transmitter and control device specification sheets have been covered and the *Remarks* section of each should be used to detail connection to a control system, including detail of the system type, signal standards, etc. The installation methods employed for these devices have already been covered in the appropriate sections and should be fully observed to ensure good, reliable control, no matter which control system is employed. Most signal converters also need to be specified as isolated to avoid earth loop problems, and can themselves be used to isolate signals from non-isolated devices for connection to the control system. Where there is a possibility of voltages outside the range of the Control System I/O specification a signal converter should be employed to protect the control system.

10.17   Installations in hazardous areas

Any application in an area classified as hazardous will require the applicable regulations and guidelines to be fully observed. In the case of some control systems the necessary barriers, etc. can be supplied as part of the system, but for others additional equipment, approved housings, etc. will be necessary. Before inclusion of in-line items, such as barriers, ensure that such equipment is suitable to work with the selected control system. Most control system suppliers will be able to provide a list of suitable devices, or give a specification to be applied for the selection of any devices. Full

Fig. 10.2 Signal isolator by Camille Bauer Controls Ltd.

details of all classified areas must be provided in all enquiry documentation to system suppliers, and to any installation contractor. If this information is not specified clearly, neither the supplier, nor installer, can be held responsible for providing the correct equipment for the application.

11 Engineering check-out to commissioning

N.B. *Engineering check-out should only be carried out by competent, qualified personnel.*

Having installed the control system we now need to carry out the **Engineering check-out** *before any power is applied* to the control system. As the title suggests this is a physical check of the installation and wiring-up of the control system. Some sites will already have a Standard applicable to this task which should be followed. The following are proposed as guidelines where such Standards do not exist, as a complement where Standards may not cater for PLCs or DCS control systems. If a site Standard does not exist for engineering check-out procedures it will be worth drawing up a simple check sheet for each loop/system I/O point. Examples of such check-out sheets for control loop equipment, analogue I/O for both PLC and DCS, and digital/analogue cabling are included at the end of this chapter. There can be a sheet per loop, or just sheets covering several system I/O per sheet. On many sites it is essential to document the engineering checks carried out prior to commissioning, and getting into the habit will make compliance with such procedures easy.

11.1 Before turning on the power

Motor control centres, etc. should not be powered at this stage and each motor should be 'locked off'.

Reference should be made to the manufacturer's specifications to ensure that all requirements of the positioning of equipment are correct, and a final check should ensure that the correct cabinetry, operator terminals, etc. has been supplied, and the various components of the system are in their correct locations.

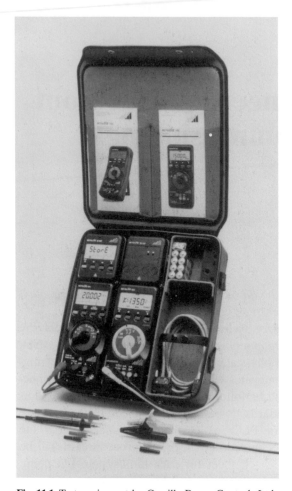

Fig. 11.1 Test equipment by Camille Bauer Controls Ltd.

¥ One person should be nominated to plan and co-ordinate all engineering check-outs.

¥ Firstly check that all 'shipping stops' have been removed from all control equipment, including measurement and control devices.

¥ The communications highway should be physically checked for correct connection, etc. at all 'nodes'/'taps', etc.

¥ If applicable ensure all I/O cards are located in the correct bins and slots, and are in a disconnected state, e.g. pulled forward.

¥ All applicable hazardous area equipment is correctly installed.

¥ All measurement devices are installed correctly, in all respects!

¥ All control devices are installed correctly, in all respects!

¥ All other field equipment is installed correctly, in all respects!

¥ With help of an assistant using a portable signal injector simulating an analogue transmitter, inject a signal in place of each transmitter in turn, confirming cabling continuity at the control system input terminals and

Fig. 11.2 Test equipment by Camille Bauer Controls Ltd.

correct I/O connection, with a test meter across the input terminals. Each of the cable cores should also be checked for 'freedom from earth', and screens for 'earth connection'.

¥ Repeat the above in place of the outputs from the control system to the control devices.

¥ Repeat the above for each contact input and output (the same injector can be used).

¥ The earthing of the complete control system must be fully checked and tested to ensure that all items requiring earth are connected, and that the earth is 'clean'.

¥ The power supplies should be checked, with control system disconnected or isolated, for correct connection and voltage. *With dc supplies, especially, check polarity before final connection to any part of the control system.*

Before continuing any further check whether the manufacturer's representative is responsible for powering up the system. In most cases initial powering up must be done by the manufacturer to comply with the terms and conditions of the control system guarantee. Even if this is not the case it is better to have the manufacturer's representative present to observe that correct procedures are followed and witness first hand any faults occurring during the powering up of the control system. With many manufacturers the procedure for powering up will repeat many of the checks detailed above, but time will be saved and embarrassment avoided if these items have already been confirmed to be correct before the representative is on site. Having satisfied all the above checks, and corrected any faults found, the control system can be powered up.

11.2   Turn on the power

Following successful powering up of the control system the programme/configuration should next be loaded onto the system, if applicable. The displays for each terminal should now be confirmed to be correct. If the control system does not require to be turned off for replacement and/or removal of I/O cards, the I/O cards can now be pushed home, one at a time, confirming each to be functioning correctly whilst doing so. Any reversed transmitter connections can be identified at this stage *and should be corrected now*. If the control system must be turned off for card removal and replacement, the best procedure is to locate all the I/O cards in one bin, or cabinet, and check the I/O, on turning on the power, for reversal, etc. Once all I/O have been confirmed correct and functional on the powered control system, all displays are correct, and showing a 'process stopped' condition, then the control system can be confirmed 'ready for pre-commissioning'.

When powering up the control system ensure that no personnel are liable to have hands, fingers, etc. in any device connected to the control system which may have high voltage applied, or may move.

11.3   Pre-commissioning the control system

Motor control centres, etc. should not be powered at this stage, and each motor should be 'locked off'.

- ¥ Again one person should be nominated to plan and co-ordinate pre-commissioning.
- ¥ Each control system analogue input should be simulated with the portable signal injector to confirm the displays give the correct information, alarms, etc.
- ¥ With each control system analogue output in manual each control device should be run through its range to confirm correct response to the control system output. Ensure personnel are *safe* from any movement of control devices, etc.
- ¥ Each control system contact input should be simulated to confirm correct display, etc.
- ¥ Each control system contact output should be put in manual, switched and confirmed correct.
- ¥ Any errors found in the above tests must be corrected before proceeding further.

¥ At this stage, if all tests have proved correct, the control system can be handed over to the plant commissioning team for the 'process commissioning' to proceed.

11.4 Commissioning the control system/ pre-commission plant

This task will normally be combined with pre-commissioning of the process plant, e.g. washing out of the plant prior to commissioning, and will generally be the overall responsibility of others. Removal of **motor control centre lockouts**, etc. should be classed as part of 'process plant commissioning', main lockouts only being removed by authorised personnel following removal of 'tradesmen's personal locks'. It is worth pointing out here that a considerable amount of time can be saved by thorough engineering check-out, and pre-commissioning, prior to the process commissioning starting. Proving all controls, motor rotations, etc. are correct will save trying to carry this out under the pressure of a plant start-up. It is at such times as these when accidents are likely to happen, and any effort spent which reduces this risk is very worthwhile.

11.5 Tuning the controllers

The process will most likely be started up initially under manual control during commissioning, to enable the control loops to be **tuned** individually, and each piece of equipment to be fully checked out under running conditions. The tuning and selection of control action of controllers which should have been done during engineering check-out, is rather specialised and should be left in the hands of qualified personnel to undertake, e.g. instrument technicians, etc. On many plants the loop tuning may take several days to finalise, since the changes of process throughput, etc. can affect the necessary tuning parameters required to a large degree, resulting in repeated re-tuning of control loops in many cases. Where cascade control, and other multiple loop control strategy, is employed the tuning of all the associated controllers may take many hours. Some control systems have the facility available of **automatic tuning**, but again this should only be used by fully qualified and trained personnel since some process conditions can cause incorrect tuning by this method, resulting in process instability under some conditions, or at worst damage to plant. Once a control loop has been tuned it is unlikely that the tuning will require further adjustment unless the process characteristics change significantly. It is generally the case that if instability is shown by a controller there is most likely to be a cause outside the capability of the control loop, e.g. a pressure controller which shows repeated rapid rise and fall of the measured variable may be showing a safety valve lifting. Equally the same symptoms on a consistency control loop may be showing a blocked stock pump.

As a guide most processes will run adequately with Proportional plus Integral control actions. Most manufacturers of electronic controllers provide default settings of Proportional and Integral actions which will give a reasonable level of automatic control on a high number of applications. As stated above, the tuning of controllers is a specialised job, and should be entrusted to an instrument technician, or personnel trained to tune control systems. Badly tuned controllers do not only generate process instability, but can cause dangerous situations to develop. *Please leave this task to the specialists!*

11.6   Troubleshooting

Some control system I/O have voltage limitation circuits built in, especially on 4 to 20 mA analogue inputs. Where direct powered transmitters are employed, e.g. magnetic flowmeters, weighing systems, etc., the voltage generated by the transmitter may be in excess of the allowable voltage presented to the control system, causing the protection system to operate. An alarm may be generated, but often the input will be seen to be zero, or open circuit. If this occurs contact the transmitter supplier to establish if the output voltage can be reduced, or a suitable zener diode permanently applied across the input terminals of the control system to keep the voltage within limits. Generally if this sort of problem occurs the assistance of a qualified instrument technician should be sought.

Following successful commissioning of process plant in many cases problems with the control of the process are initially blamed on the control system. This is primarily because the control system is the 'window' through which the process is observed, and any fault, whether control or process, is displayed on the control equipment. Providing the engineering check-out and commissioning have been carried out thoroughly, the control system is most likely to show a problem in the process, or possibly the tuning of the control loop. With most DCS available today, a high degree of **self-diagnostics** is inbuilt into the system, giving display and alarms of any hardware faults, along with many software faults occurring within the control system itself. With some PLC systems and single-loop controllers, a degree of self-diagnostics is available, but generally to a much reduced degree in comparison to the majority of DCS, and normally only when video terminals/ programming units are connected, utilising sophisticated SCADA software. In most cases of faults outside the scope of self-diagnostics, the error is generally outside of the control system, except for tuning problems. In such situations the control loop should be put into manual control mode, the controller output set to a known operable value, and the process parameter closely observed to establish if the variable at fault regains stability. The following conditions will give an indication of the fault.

1. If stability is not regained within a short time period the control device should be checked to ensure that it is responding to the control signal correctly.
2. If all is well with the control device and instability continues, then the problem is more serious and should be quickly passed to a qualified process technician to sort out, since it is either process instability, interaction from another control loop or a failure of a piece of process plant, etc.
3. If stability is regained the problem is most likely to be associated with the control loop tuning, over-ranged measurement device or oversized/undersized control valve, etc. (see Chapter 3), and a qualified technician should be called. (Over-ranged devices should be apparent on newly commissioned plant, and not on plant that has been operating for some time.)
4. If modulating the control device has no effect on the measured parameter the measurement device and its installation should be thoroughly examined, including a check on all isolating valves to ensure that they are in their correct positions.
5. If stability is regained for a short period and then instability re-occurs the indication is that another part of the process is influencing the loop in trouble. In such cases, if the cause is not immediately obvious, a qualified technician or process engineer should be called. Re-tuning the affected control loop may overcome the problem for only a short time, or until the plant throughput, etc. is changed.

Programmes and configuration do not change on their own, and if the process has run in a stable, controlled state for some time, the programme or configuration will rarely be at fault. Sometimes, however, communications on networked systems can become overloaded in certain situations. Observe the time response to update displays, and reaction time to changes in set points, etc. If these response times change, or are excessively long, there could possibly be a problem with the amount of data being transferred over the communications system at certain times. If this is suspected a qualified technician should be called to rectify the problem.

Remember: when changing any programme or configuration, upgrading software levels, etc. always ensure a full copy of the old, working, programme, etc. is made before the changes are installed. If the new software, programme, etc. does not function correctly you can go back to the one that did by re-installing the copy of the working programme. This may sound a little obvious, but you would be surprised how many times this is not done, usually with very embarrassing, or even catastrophic, results!

Hopefully the above procedures will identify the possible cause of the trouble, and enable a solution to be put in place, allowing the commissioning/running of the process and plant to continue without further problems.

At this stage the process control aspect of the project should be deemed a great success, and the experience gained applied to pave the way for the next successful application of process control. *But remember, each project is different, and do not forget: you get what you pay for*, and *if it is not specified, it will not be included.*

This is not the final stage since current legislation, as well as maintenance efficiency, requires all process plant to be fully documented. Hopefully the project has generated a complete set of documentation for all the equipment, how it has been installed, programmed, etc. Unfortunately not all projects go exactly to plan, or design, and slight modifications are necessary during commissioning, and the early stages of getting the plant into full production. These modifications must be fully documented and all changes detailed as well as drawings, etc. updated to reflect the '**as installed and operational**' state of the equipment. Only after this has been done can the project be deemed completed.

PROCESS CONTROL
ENGINEERING CHECK-OUT SHEET
Digital/Analogue Cabling

I/O No.	Type	Tag No	Cable No	Cores	Test Result	Initials

ACCEPTANCE / APPROVAL

ENGINEER:-   Date:-

PROCESS CONTROL
ENGINEERING CHECK-OUT SHEET
Analogue

SERVICE: LOOP TAG No.:

DRAWING No.:

MEASUREMENT/INPUT

	DETAILS		Fail	Pass
Tx Type				
Mounting				
Housing				
Process Connection				
Signal Connection	Tx	Input		
Cable/Tube Ident	Tx	Input		
Cores	Tx	Input		
Signal Injected				
Result				

ENGINEER: Date:

CONTROL DEVICE/OUTPUT

	DETAILS		Fail	Pass
Device Type				
Mounting				
Housing				
Process Connection				
Signal Connection	Device	Output		
Cable/Tube Ident	Device	Output		
Cores	Device	Output		
Signal Injected				
Result				

ENGINEER: Date:

PROCESS CONTROL
ENGINEERING CHECK-OUT SHEET

LOOP TAG No.:

SERVICE:

MEASUREMENT DEVICE TYPE:

MOUNTING:

HOUSING:

PROCESS CONNECTION:

SIGNAL CONNECTION:

SIGNAL CABLE/TUBE IDENT.:

CORE IDENTS.:

SIGNAL INJECTED:

RESULT:

CONTROL DEVICE TYPE:

MOUNTING:

HOUSING:

PROCESS CONNECTION:

SIGNAL CONNECTION:

SIGNAL CABLE/TUBE IDENT.:

CORE IDENTS.:

SIGNAL INJECTED:

RESULT:

COMMENTS:

Appendix: Conversion tables

AREA

YOU WANT

YOU HAVE	Square Miles	Morgen	Acres	Square Yards	Square Feet	Square Kilometres	Hectares	Ares	Square Metres
Square Miles	1	302.3818	640	3097.6×10^3	27878.4×10^3	2.58999	258.999	25899.9	2589.99×10^3
Morgen	3.3071×10^{-3}	1	2.11653	10244.0524	92196.4717	8565.32×10^{-6}	856.532×10^{-3}	85.6532	8565.32
Acres	1.5625×10^{-3}	472.471×10^{-3}	1	4840	43560	4046.86×10^{-6}	404.686×10^{-3}	40.4686	4046.86
Square Yards	0.32283×10^{-6}	97.618×10^{-6}	206.612×10^{-6}	1	9	836.127×10^{-9}	83.6127×10^{-6}	8.36127×10^{-3}	0.836127
Square Feet	35.8701×10^{-9}	10.8464×10^{-6}	22.95684×10^{-6}	0.111111	1	92.903×10^{-9}	9.2903×10^{-6}	929.03×10^{-6}	92.903×10^{-3}
Square Kilometres	386.102×10^{-3}	116.7499	247.105	1195.99×10^3	10.76391×10^6	1	100	10000	10^6
Hectares	3.86102×10^{-3}	1.167499	2.47105	11959.9	107.6391×10^3	0.01	1	100	10000
Ares	38.6102×10^{-6}	11.67499×10^{-3}	24.7105×10^{-3}	119.599	1076.391	0.1×10^{-3}	0.01	1	100
Square Metres	386.102×10^{-9}	116.7499×10^{-6}	247.105×10^{-6}	1.19599	10.7639	1.0×10^{-6}	0.1×10^{-3}	0.01	1

AREA – 2

		YOU WANT				
	Square Yards	Square Feet	Square Inches	Square Metres	Square Centimetres	Square Millimetres
Square Yards	1	9	1296	0.836127	8.36127×10^3	836.127×10^3
Square Feet	0.111111	1	144	92.903×10^{-3}	929.03	92903
Square Inches	0.771605×10^{-3}	6.94444×10^{-3}	1	0.64516×10^{-3}	6.4516	645.16
Square Metres	1.19599	10.7639	1550	1	10×10^3	10^6
Square Centimetres	0.11959×10^{-3}	1.07639×10^{-3}	0.155	0.1×10^{-3}	1	100
Square Millimetres	1.19599×10^{-6}	10.7639×10^{-6}	1.55×10^{-3}	10^{-6}	0.01	1

YOU HAVE

FORCE

YOU WANT

YOU HAVE

	Mega Newtons	Kilo Newtons	Newtons	Kg Force	Tonnes Force	Short Tons Force	Long Tons Force	Pounds Force	Ounces Force
Mega Newtons	1	10^3	10^6	101.972×10^3	101.972	112.405	100.361	224.809×10^3	3.596944×10^6
Kilo Newtons	10^{-3}	1	10^3	101.972	101.972×10^{-3}	112.405×10^{-3}	100.361×10^{-3}	224.809	3.596944×10^3
Newtons	10^{-6}	10^{-3}	1	101.972×10^{-3}	101.972×10^{-6}	112.405×10^{-6}	100.361×10^{-6}	224.809×10^{-3}	3.596944
Kg Force	9.80665×10^{-6}	9.80665×10^{-3}	9.80665	1	10^{-3}	1.10231×10^{-3}	0.984206×10^{-3}	2.20462	35.27392
Tonnes Force	9.80665×10^{-3}	9.80665	9.80665×10^3	10^3	1	1.10231	0.984206	2204.62	35.27392×10^3
Short Tons Force	8.89644×10^{-3}	8.89644	8.89644×10^3	907.184	907.184×10^{-3}	1	0.892858	2000	32000
Long Tons Force	9.96402×10^{-3}	9.96402	9.96402×10^3	1016.05	1.01605	1.12	1	2240	35840
Pounds Force	4.44822×10^{-6}	4.44822×10^{-3}	4.44822	453.592×10^{-3}	453.592×10^{-6}	0.5×10^{-3}	0.446429×10^{-3}	1	16
Ounces Force			278.01378×10^{-3}	28.3495×10^{-3}	28.3495×10^{-6}	31.25×10^{-6}	27.90178×10^{-6}	0.0625	1

DENSITY

<table>
<tr><td rowspan="2"></td><td colspan="6" align="center">YOU WANT</td></tr>
<tr><td>Grammes/cm^3</td><td>Kg/m^3</td><td>Tonnes/m^3</td><td>Lbs/in^3</td><td>Lbs/ft^3</td><td>Short Tons/yd^3</td></tr>
<tr><td>Grammes/cm^3</td><td>1</td><td>10^3</td><td>1</td><td>36.1273×10^{-3}</td><td>62.428</td><td>842.77×10^{-3}</td></tr>
<tr><td>Kg/m^3</td><td>10^{-3}</td><td>1</td><td>10^{-3}</td><td>36.1273×10^{-6}</td><td>62.428×10^{-3}</td><td>842.77×10^{-6}</td></tr>
<tr><td>Tonnes/m^3</td><td>1</td><td>10^3</td><td>1</td><td>36.1273×10^{-3}</td><td>62.428</td><td>842.77×10^{-3}</td></tr>
<tr><td>Lbs/in^3</td><td>27.6799</td><td>27.6799×10^3</td><td>27.6799</td><td>1</td><td>1728</td><td>23.328</td></tr>
<tr><td>Lbs/ft^3</td><td>16.0185×10^{-3}</td><td>16.0185</td><td>16.0185×10^{-3}</td><td>0.57870×10^{-3}</td><td>1</td><td>13.5×10^{-3}</td></tr>
<tr><td>Short Tons/yd^3</td><td>1.18656</td><td>1.18656×10^3</td><td>1.18656</td><td>42.8669×10^{-3}</td><td>74.0741</td><td>1</td></tr>
</table>

Y O U

H A V E

LINEAR

						YOU WANT				
		Miles	Yards	Feet	Inches	Kilometres Km	Metres M	Centimetres cm	Millimetres mm	Cape Feet
YOU	Miles	1	1760	5280	63360	1.609344	1609.344	160.9344×10^3	1609.344×10^3	5111.3304
	Yards	568.1818×10^{-6}	1	3	36	0.9144×10^{-3}	0.9144	91.44	914.4	2.904165
	Feet	189.39393×10^{-6}	0.333333	1	12	0.3048×10^{-3}	0.3048	30.48	304.8	0.968055
HAVE	Inches	15.78283×10^{-6}	27.7778×10^{-3}	83.333×10^{-3}	1	0.0254×10^{-3}	0.0254	2.54	25.4	0.080671
	Kilometres Km	0.62137	1093.61	3280.84	39370.1	1	1000	100×10^3	10^6	3176.034
	Metres M	0.62137×10^{-3}	1.09361	3.28084	39.3701	10^{-3}	1	100	10^3	3.176034
	Centimetres cm	6.2137×10^{-6}	10.9361×10^{-3}	32.8084×10^{-3}	0.393701	0.01×10^{-3}	0.01	1	10	31.76034×10^{-3}
	Millimetres mm	0.62137×10^{-6}	1.09361×10^{-3}	3.28084×10^{-3}	39.3701×10^{-3}	10^{-6}	10^{-3}	0.1	1	3176.034×10^{-6}
	Cape Feet	195.644×10^{-6}	0.344333	1032.9999×10^{-3}	12.395888	0.314858×10^{-3}	0.314858	314.4858	314.8581	1

MASS

YOU WANT (columns) — **YOU HAVE** (rows)

YOU HAVE \ YOU WANT	Tonnes	Kg	Grammes	Short Tons	Long Tons	Kips	Pounds	Ounces	Grains
Tonnes	1	1000	10^6	1.10231	0.984205	2.20462	2204.52	35.2739×10^3	15.43234×10^6
Kg	0.001	1	10^3	1.10231×10^{-3}	984.205×10^{-6}	2.20462×10^{-3}	2.20462	35.2739	15.43234×10^3
Grammes	10^{-6}	0.001	1	1.10231×10^{-6}	984.205×10^{-9}	2.20462×10^{-6}	2.20462×10^{-3}	35.2739×10^{-3}	15.43234
Short Tons	907.185×10^{-3}	907.185	907.185×10^3	1	0.892857	2	2000	32000	
Long Tons	1.016048	1016.048	1016.048×10^3	1.12	1	2.24	2240	35840	
Kips	453.59237×10^{-3}	453.59237	453.59237×10^3	0.5	0.446423	1	1000	16000	
Pounds	453.59237×10^{-6}	453.59237×10^{-3}	453.59237	0.5×10^{-3}	0.446428×10^{-3}	0.001	1	16	7000
Ounces	28.3495×10^{-6}	28.3495×10^{-3}	28.3495	31.25×10^{-6}	27.90178×10^{-6}	62.5×10^{-6}	0.0625	1	437.5
Grains	64.798×10^{-9}	64.7989×10^{-6}	64.7989×10^{-3}				0.14285×10^{-3}	2.28571×10^{-3}	1

POWER

YOU HAVE \ YOU WANT	Giga Watts	Mega Watts	Kilo Watts	Watts	M.Kg force/Sec.	Ft.Pound force/Sec.	Ft.Pound force/Min.	Metric H.P.	H.P.
Giga Watts	1	10^{3}	10^{6}	10^{9}	0.101972×10^{9}	0.737562×10^{9}	44.25372×10^{9}	1.35962×10^{6}	1.34102×10^{6}
Mega Watts	10^{-3}	1	10^{3}	10^{6}	0.101972×10^{6}	0.737562×10^{6}	44.25372×10^{6}	1.35962×10^{3}	1.34102×10^{3}
Kilo Watts	10^{-6}	10^{-3}	1	10^{3}	0.010972×10^{3}	0.737562×10^{3}	44.25372×10^{3}	1.35962	1.34102
Watts	10^{-9}	10^{-6}	10^{-3}	1	0.101972	0.737562	44.25372	1.35962×10^{-3}	1.34102×10^{-3}
M.Kg force/Sec.	9.80665×10^{-9}	9.80665×10^{-6}	9.80665×10^{-3}	9.80565	1	7.23301	433.9806	13.3333×10^{-3}	13.1509×10^{-3}
Ft.Pound force/Sec.	1.35582×10^{-9}	1.35582×10^{-6}	1.35582×10^{-3}	1.35582	0.138255	1	60	1.8434×10^{-3}	1.81818×10^{-3}
Ft.Pound force/Min.	22.59697×10^{-12}	22.59697×10^{-9}	22.59697×10^{-6}	22.59697×10^{-3}	2.30425×10^{-3}	16.66666×10^{-3}	1	30.72333×10^{-6}	30.30303×10^{-6}
Metric H.P.	735.499×10^{-9}	735.499×10^{-6}	735.499×10^{-3}	735.499	75	542.476	32.54856×10^{3}	1	0.98632
H.P.	745.7×10^{-9}	745.7×10^{-6}	745.7×10^{-3}	745.7	76.0402	550	33×10^{3}	1.01387	1

PRESSURE

	YOU WANT								
YOU HAVE	Atmospheres	mm Hg	mm Water	Feet Water	Inches Water	P.S.I.	Kg/cm²	N/m²	Bar
Atmospheres	1	760	10.33227×10^3	33.9007	406.8084	14.69595	1.03323	101.325×10^3	1.01325
mm Hg	1.31579×10^{-3}	1	13.5951	44.6033×10^{-3}	0.53524	19.3368×10^{-3}	1.35951×10^{-3}	133.322	1.33322×10^{-3}
mm Water	96.78415×10^{-6}	73.55591×10^{-3}	1	3.28084×10^{-3}	39.3701×10^{-3}	1.42233×10^{-3}	0.1×10^{-3}	9.80665	98.0665×10^{-6}
Feet Water	29.49791×10^{-3}	22.41985	304.8	1	12	0.43353	30.48×10^{-3}	2989.07	29.8907×10^{-3}
Inches Water	2.45816×10^{-3}	1.86832	25.4	83.333×10^{-3}	1	36.127×10^{-3}	2.54×10^{-3}	249.089	2.49089×10^{-3}
P.S.I.	68.04594×10^{-3}	51.715	703.0717	2.30666	27.6799	1	70.307×10^{-3}	6.89476×10^3	68.9476×10^{-3}
Kg/cm²	967.84152×10^{-3}	735.55913	10×10^3	32.308	393.701	14.2233	1	98.0665×10^3	980.665×10^{-3}
N/m²	9.86923×10^{-6}	7.5006×10^{-3}	101.9716×10^{-3}	334.553×10^{-6}	4.01463×10^{-3}	145.038×10^{-6}	10.1972×10^{-6}	1	10^{-6}
Bar	986.923×10^{-3}	750.062	10.19716×10^3	33.4553	401.463	14.5038	1.01972	100×10^3	1

VELOCITY

<table>
<tr><td rowspan="2"></td><td rowspan="2"></td><td colspan="5" align="center">YOU WANT</td></tr>
</table>

YOU HAVE	Km/Hour	Metres/second	Miles/Hour	Feet/minute	Feet/second
Km/Hour	1	0.277778	0.621371	54.68064	0.911344
Metres/second	3.6	1	2.23694	196.8504	3.28084
Miles/Hour	1.609344	0.44704	1	88	1.46667
Feet/minute	0.018288	0.00508	11.3636×10^{-3}	1	0.016666
Feet/second	1.09728	0.3048	0.681818	60	1

FLOW RATE

	YOU WANT									
(YOU HAVE)	Imperial G.p.m.	Imperial G.p.h.	Ft³/Sec.	Ft³/Min. c.f.m.	m³/Sec.	m³/Min.	m³/Hour	Litres/Sec.	Litres/Min.	U.S. G.p.m.
Imperial G.p.m.	1	60	2.67574×10^{-3}	180.544×10^{-3}	75.7632×10^{-6}	4.54609×10^{-3}	272.7657×10^{-3}	75.7682×10^{-3}	4.54609	1.20094
Imperial G.p.h.	16.6666×10^{-3}	1	44.595×10^{-6}	2.67574×10^{-3}	1.2628×10^{-6}	75.7682×10^{-6}	4.54609×10^{-3}	1.2628×10^{-3}	75.7682×10^{-3}	20.01569×10^{-3}
Ft³/Sec.	373.728	22.42368×10^{3}	1	60	28.31686×10^{-3}	1.69901	10194069	28.31686	1699.013	448.833
Ft³/Min. c.f.m.	6.2288	373.728	16.6666×10^{-3}	1	471.94777×10^{-6}	28.31686×10^{-3}	1.69901	471.94777×10^{-3}	28.31686	7.48052
m³/Sec.	13.19815×10^{3}	791.8884×10^{3}	35.31465	2118.88	1	60	3600	1000	60×10^{3}	15.8502×10^{3}
m³/Min.	219.969	13.19815×10^{3}	588.5775×10^{-3}	35.31465	16.6666×10^{-3}	1	60	16.66666	1000	264.17
m³/Hour	3.66615	219.969	9.80962×10^{-3}	588.5775×10^{-3}	277.7777×10^{-6}	16.6666×10^{-3}	1	277.77777×10^{-3}	16.66666	4.40288
Litres/Sec.	13.19815	791.8884	35.31465×10^{-3}	2.11888	10^{-3}	0.06	3.6	1	60	15.85017
Litres/Min.	219.969×10^{-3}	13.19815	588.577×10^{-6}	35.31465×10^{-3}	16.66666×10^{-6}	10^{-3}	0.06	16.66666×10^{-3}	1	264.17×10^{-3}
U.S. G.p.m.	0.83268	49.9608	2.228×10^{-3}	133.6805×10^{-3}	63.0907×10^{-6}	3.78544×10^{-3}	227.124×10^{-3}	63.0907×10^{-3}	3.78544	1

VOLUME

	Cubic Metres	Litres	Cubic cm	Cubic Yards	Cubic Feet	Cubic Inches	Imperial Gallons	Pints	Fluid Ounces	US Gallons
Cubic Metres	1	1000	10^6	1.30795	35.3147	61.0236×10^3	219.969	1759.754	35.1951×10^3	264.171
Litres	10^{-3}	1	1000	1.30795×10^{-3}	35.3147×10^{-3}	61.0236	219.969×10^{-3}	1759.754×10^{-3}	35.1951	264.171×10^{-3}
Cubic cm	10^{-6}	10^{-3}	1	1.30795×10^{-6}	35.3147×10^{-6}	61.0236×10^{-3}	219.969×10^{-6}	1759.754×10^{-6}	35.1951×10^{-3}	264.171×10^{-6}
Cubic Yards	0.76455	764.55	764.55×10^3	1	27	46.656×10^3	168.178	1354.424	26.90848×10^3	201.974
Cubic Feet	28.3168×10^{-3}	28.3168	28.3168×10^3	37.037×10^{-3}	1	1728	6.2288	49.8304	996.608	7.48052
Cubic Inches	16.3871×10^{-6}	16.3871×10^{-3}	16.3871	21.43347×10^{-6}	0.5787×10^{-3}	1	3.60464×10^{-3}	28.8371×10^{-3}	0.57674	4.329×10^{-3}
Imperial Gallons	4.54609×10^{-3}	4.54609	4.54609	5.9461×10^{-3}	0.160544	277.42003	1	8	160	1.20094
Pints	568.26125×10^{-6}	568.26125×10^{-3}	568.26125	0.74326×10^{-3}	20.068×10^{-3}	34.6775	0.125	1	20	150.1176×10^{-3}
Fluid Ounces	28.413×10^{-6}	28.413×10^{-3}	28.413	37.163×10^{-6}	1.0034×10^{-3}	1.73388	6.25×10^{-3}	0.05	1	7.50583×10^{-3}
US Gallons	3.7854×10^{-3}	3.7854	3.7854×10^3	4.9513×10^{-3}	133.6805×10^{-3}	231	0.83268	6.66144	133.2288	1

YOU HAVE

	YOU WANT						
	MN/m²	KN/m²	N/m²	Kg Force/cm²	Pound Force/in²	Short Tons Force/in²	Long Tons Force/in²
Y O U H A V E MN/m²	1	10^3	10^6	10.1972	145.037783	72.51889×10^{-3}	64.749×10^{-3}
KN/m²	10^{-3}	1	10^3	10.1972×10^{-3}	$145.037783 \times 10^{-3}$	72.51889×10^{-6}	64.749×10^{-6}
N/m²	10^{-6}	10^{-3}	1	10.1972×10^{-6}	$145.037783 \times 10^{-6}$	72.51889×10^{-9}	64.749×10^{-9}
Kg Force/cm²	98.0665×10^{-3}	98.0665	98.0665×10^3	1	14.2233	7.11167×10^{-3}	6.34971×10^{-3}
Pound Force/in²	6.89476×10^{-3}	6.89476	6.89476×10^3	70.307×10^{-3}	1	0.5×10^{-3}	446.429×10^{-6}
Short Tons Force/in²	13.7895	13.7895×10^3	13.7895×10^6	140.614	2000	1	0.892858
Long Tons Force/in²	15.4443	15.4443×10^3	15.4443×10^6	157.488	2240	1.12	1

Index